PRICING WATER RESOURCES AND WATER AND SANITATION SERVICES

OECD

ORGANISATION FOR ECONOMIC CO-OPERATION AND DEVELOPMENT

The OECD is a unique forum where the governments of 30 democracies work together to address the economic, social and environmental challenges of globalisation. The OECD is also at the forefront of efforts to understand and to help governments respond to new developments and concerns, such as corporate governance, the information economy and the challenges of an ageing population. The Organisation provides a setting where governments can compare policy experiences, seek answers to common problems, identify good practice and work to co-ordinate domestic and international policies.

The OECD member countries are: Australia, Austria, Belgium, Canada, the Czech Republic, Denmark, Finland, France, Germany, Greece, Hungary, Iceland, Ireland, Italy, Japan, Korea, Luxembourg, Mexico, the Netherlands, New Zealand, Norway, Poland, Portugal, the Slovak Republic, Spain, Sweden, Switzerland, Turkey, the United Kingdom and the United States. The Commission of the European Communities takes part in the work of the OECD.

OECD Publishing disseminates widely the results of the Organisation's statistics gathering and research on economic, social and environmental issues, as well as the conventions, guidelines and standards agreed by its members.

This work is published on the responsibility of the Secretary-General of the OECD. The opinions expressed and arguments employed herein do not necessarily reflect the official views of the Organisation or of the governments of its member countries.

ISBN 978-92-64-08346-2 (print)
ISBN 978-92-64-08360-8 (PDF)

Also available in French: *Le prix de l'eau et des services d'eau potable et d'assainissement*

Photo credits: Cover:
©Taro Yamada/Corbis, © Yellowj/Dreamstime.com, © Bart Sadowski/iStockphoto, © jerome Moreaux/Fotolia.com

Corrigenda to OECD publications may be found on line at: *www.oecd.org/publishing/corrigenda*.

Foreword

This report is one of the outputs of the OECD's 2007-08 Horizontal Water Programme, which aimed to address two sets of questions:

- how to overcome the financial obstacles to the provision of adequate and affordable water and sanitation services that are affordable to all, including the least well off, while ensuring the financial sustainability of service providers; and
- how to improve the use of economic incentives to encourage management of water resources that is both economically efficient and environmentally sustainable, with a special focus on water use in agriculture.

This report presents the main findings and policy conclusions regarding the use of pricing instruments for water resources management and water and sanitation services.

An annotated outline, conceptual framework and intermediary results were discussed at the OECD Working Party on Global and Structural Policies (WPGSP) in 2008. The report builds on the comments and suggestions received from WPGSP delegates and from discussions at a number of conferences, including an expert meeting held in Paris in November 2007, the 2008 Stockholm World Water Week, the Global Forum on Sustainable Development held at OECD headquarters in Paris on 1-2 December 2008, and the 5th World Water Forum in Istanbul in March 2009.

The first draft of the report has been prepared by Monica Scatasta, who was seconded to OECD in the Environment Directorate from the European Investment Bank for the duration of the 2007-08 Programme. Work was carried out under the supervision of Brendan Gillespie. The author has worked with support from Carla Bertuzzi, Virginia Dagostino and David Kimble. In addition, input on different aspects of the report was received from three consultants (in alphabetical order): Paul Herrington, Antonio Massarutto and Robin Simpson. The final draft was developed by Xavier Leflaive, under the supervision of Anthony Cox. Colleagues from the Environment Directorate and Henri Smets provided additional comments.

Table of Contents

Tables

Figures

Boxes

Acronyms

3Ts	Tariffs, taxes, transfers
ABH	Basin Water Authorities (Portugal)
BGW	German Water Association
BOD	Biochemical oxygen demand
COD	Chemical oxygen demand
CVD	*Coût vérité distribution*
CPI	Consumer price index
EAP	Environmental action plan
EEA	European Environment Agency
EECCA	Eastern Europe, the Caucasus and Central Asia
FCR	Full cost recovery
GDP	Gross domestic product
GWI	Global Water Intelligence
IBT	Increasing-block tariff
INAG	National Water Authority in charge of water resources management (Portugal)
IRAR	Economic regulator for water supply and sanitation (Portugal) (ERSAR, as of October 2009)
LBL	Load-based licensing
MDGs	Millennium Development Goals
OECD	Organisation for Economic Development and Co-operation
MC	Marginal cost
O&M	Operation and maintenance
p.e.	population equivalent
PPP	Purchasing power parity
SCR	Sustainable cost recovery
TSS	Total suspended solids
UAE	United Arab Emirates
UNESCO	United Nations Educational, Scientific and Cultural Organization
VAT	Value-added tax
WHO	World Health Organization
WPGSP	Working Party on Global and Structural Policies
WRM	Water resource management
WSS	Water supply and sanitation

Executive summary

Objective and scope of the report

This report updates two previous reviews of OECD countries' experience with pricing for water-related services. It is based on the OECD 2007-08 Survey which was set up to address two sets of policy questions:

- What proportion of costs is recovered by revenues from tariffs? What alternative cost-sharing mechanisms may be appropriate for different water-related activities?
- Are average tariff levels affordable for all? And if not, are adequate tariff structures, including social tariffs, or other income-support mechanisms being adopted?

The data was collected at national and local levels, on prices and tariff structures for water supply and sanitation (WSS) services, for domestic (households) and industrial uses. It also covers the diffusion of metering, and cost recovery. In addition, available information on abstraction and pollution charges is presented. Data was used to assess affordability of water supply and sanitation in OECD countries.

A new context for water pricing

Two main sets of challenges face the water sector in OECD and non-OECD countries. One regards the increasing competition for the use of water resources for human consumption, productive uses, and the support of ecosystems. Limited availability of water resources, their deteriorating quality, the impacts of climate change and poor management, all contribute to the problem.

The other challenge is the need to ensure access to adequate, sustainable and affordable water and sanitation services for all, including poor households. While in some regions water scarcity is one limiting factor, this goal is mainly constrained by management factors including ill-conceived investment, the deterioration of infrastructure due to insufficient cash flows, or inappropriate regulatory frameworks.

These challenges are not confined to developing countries: recent analyses confirm that OECD countries are facing similar (although different) issues.

In this context, water policies need a mechanism to allocate water where it is most needed, and a financing instrument to generate revenues. Appropriately designed and tailored to local conditions, pricing water-related services can contribute to the achievement of these policies. This requires that both tariff levels and structures are considered, in co-ordination with other financing instruments (taxes and transfers).

Water pricing in OECD countries

OECD countries are gaining experience with abstraction charges, pollution/effluent charges and other economic instruments – such as tradable water use permits – to achieve more economically efficient and environmentally sustainable abstraction and allocation among competing uses. In most countries, abstraction charges are designed with the objective of providing funding for water resources management or for watershed protection activities. Despite this, they tend to be relatively low. In the limited sample available, higher charges are imposed on groundwater than on surface water. In most cases, charges are collected and retained locally.

More countries reported the use of pollution charges. Pollution charges can be linked to different characteristics of the polluter, the effluents or the recipient water body. In most cases, fees/charges are collected at the local level – but only seldom at the river basin level – to finance environmental activities. Some systems provide incentives to continuously reduce discharges in water bodies. In some instances, countries adopted alternative cost-allocation mechanisms that recognise the existence of a broader set of beneficiaries. For example, revenues collected from downstream beneficiaries are used to compensate upstream residents for losses due to land use regulation, an important step towards truly integrated water and land management at the river basin level.

Tariff levels charged to households for WSS services vary greatly among OECD countries, reflecting contrasted efforts to recover the costs of the services through prices. The data show that in half of the countries, wastewater services can be more expensive than drinking water supply. It also confirms that prices have risen over the last decade (although at times more slowly in the most recent years), primarily driven by wastewater charges, which were brought in line with the costs of investment needed for environmental compliance. Value-added tax (VAT) and other taxes also explain part of the increase.

Tariff structures for water supply vary within and across OECD countries. The diversity of tariff structures in a country reflects the degree of decentralisation of the tariff-setting process. The main difference with previous surveys is a smaller number of countries where the use of flat fees and decreasing block tariff structures were detected. An emerging trend in some OECD countries is the increasing use of fixed charges alongside volumetric components, or the progressive increase in the weight of fixed charges in the overall bill.

Increasingly, separate wastewater charges are being introduced to recover wastewater management costs. Most countries levy separate charges for sewerage *vs.* wastewater treatment, although in most cases the basis for charging remains water consumption; only the size of the volumetric rate differs.

Data collection is even more difficult for industrial water supply and sanitation services; differences across productive sectors, for instance, provide an additional layer of complexity. With regard to water supply, the main difference with household tariff structures is that a few more countries and regions use decreasing block tariffs, particularly for large users. The objective of keeping large customers that provide substantial revenues and stable flows seems to inhibit the use of tariff structures that may provide incentives to reduce water use. With regards to wastewater management, data shows an increasing use of separate charges for wastewater collection and for wastewater treatment, with the latter increasingly based on the pollution load of industrial effluents, thus better reflecting actual treatment costs.

Taxes are applied to water bills more often than a decade ago. It is noteworthy that VAT and other taxes can affect final demand and the affordability of service, but do not contribute to cost recovery. Data shows that taxes on water-related services vary greatly across countries and make cross-country comparisons difficult.

Data on water supply and sanitation in non-OECD countries was only collected at local level and national computations are difficult. Reported data indicate a (sometimes steep) increase in prices over the last decade for water supply and sanitation services, however from usually low levels. Some countries in Asia, Latin America and the Middle East have tariffs above USD 1/m^3 (compared with tariffs ranging from USD 1 to USD 4/m3 in most OECD countries). However, in most cases, tariffs provide little incentives to use water efficiently (including by curbing down leakages) and contribute little to cost recovery.

Consequences for selected policy issues

The data collected through the OECD 2007-08 Survey sheds some light on selected policy issues related to pricing water supply and sanitation services.

First, it confirms that metering is unevenly spread across OECD countries. This forbids use of the first best option for achieving economic efficiency, which is marginal cost pricing. There may be good reasons for this (metering is costly and applying marginal cost pricing to water generates difficulties), but it follows that economic efficiency can only be met through second best choices.

Second, data indicate that, in OECD countries, the operation and maintenance costs of domestic and industrial WSS services are generally covered. However, there doesn't appear to be a large margin for operators to also face the need to renew and replace ageing infrastructure, although very few countries provided data on this item. Generating revenues to cover full economic or sustainability costs seems to be a remote target only.

An analysis of specific cases suggests that efforts have been made to increase cost recovery from tariffs in many OECD countries. The focus has been primarily on ensuring that effective funding mechanisms are in place to ensure the financial sustainability of the sector, and particularly of wastewater management, where larger investments are needed.

Third, the data has made it possible to assess affordability of water supply and sanitation in selected OECD countries. Figures confirm that water supply and sanitation bills do not represent a considerable burden on disposable household income when using average income figures. The picture is more contrasted when one considers the lowest decile of the population: average representative bills would represent a significant share of the disposable income of these groups in a number of countries.

It is important to note that many countries have introduced social tariffs or accompanying measures. The analysis of these measures indicates that, properly designed and tailored to local conditions, pricing can be an effective instrument to contribute to the environmental, social, economic and financial dimensions of water policies.

There would be important benefits from more work to regularly document these trends, identify best practices (*e.g.* on financing the renewal of infrastructures or on coping with affordability issues) and learn lessons (*e.g.* on price elasticity of demand, or the effects of specific taxes). This would fill in the knowledge gaps and facilitate cross-country comparisons. The lessons could be reflected in a checklist for policy makers involved in the design, or the revision of, pricing policies for water-related services. This report paves the way forward.

Introduction

This report updates two previous reviews of the experience of OECD countries with pricing for water-related services and with the social aspects of tariffs for water supply and sanitation (see OECD 1999, 2003). It expands the geographical scope of the previous reports by including a discussion of tariffs for water supply and sanitation (WSS) in non-OECD countries. New data collected for non-OECD countries is less comprehensive than for OECD countries; it was coupled with a number of case studies.

The analysis is based on an extensive review of recent experience in OECD and selected emerging and developing countries. The main focus of the report is the level and structure of tariffs for water and sanitation services for municipal uses (domestic and industrial). Agriculture water use and pricing is addressed in a separate report issuing from the OECD Horizontal Water Programme (OECD, 2010). Regarding water resources management, this report provides a more succinct review of experiences with abstraction and pollution charges in OECD and selected non-OECD countries. A more in-depth discussion of the use of economic instruments for water resources management remains beyond the scope of this report, but will be addressed in future OECD work.

The first chapter sets the stage and explains why water pricing is a policy issue. It claims that pricing has two functions; the chapter presents the key policy questions and challenges which face policy makers in this area. Chapter 2 presents the state of play of pricing for water-related services in OECD countries. It builds on a survey of pricing practices in OECD countries; data have been updated in March 2009, on the basis of information sent to (or endorsed by) OECD countries. When needed, data from a complementary survey of water prices in selected cities from OECD and non-member countries have been used. In Chapter 3, the contribution of current practices in OECD countries to the policy challenges identified in Chapter 1 is assessed. As a conclusion, Chapter 3 identifies knowledge gaps and points to avenues for further work.

References

OECD (1999), *The Price of Water: Trends in OECD Countries*, OECD Publishing, DOI: *http://dx.doi.org/10.1787/9789264173996-en.*

OECD (2003), *Social Issues in the Provision and Pricing of Water Services*, OECD Publishing, DOI: *http://dx.doi.org/10.1787/9789264099890-en.*

OECD (2010), *Sustainable Management of Water Resources in Agriculture*, OECD Publishing, *www.oecd.org/tad/env.*

Chapter 1

Why water pricing is an issue

This chapter clarifies the multiple roles of water pricing and situates pricing among other financing mechanisms for water-related services. It identifies two policy questions which will organise the analyses in the report: What proportion of the costs of water-related services is recovered by revenues from tariffs? Are tariff structures adequate and accompanied by social mechanisms?

Pricing is an instrument that can help reconcile sound water resources management with the adequate provision of water services and investment in infrastructure. It follows that water pricing plays multiple roles. It also takes multiple dimensions: the actual price (level), the tariff structure, the accompanying measures. A fourth dimension is not be covered in this report: the process for tariff setting and revision is consequential for the three dimensions mentioned above; tariff setting for water supply and sanitation is explored in a companion report on strategic financial planning (see OECD, 2009a).

Water prices: allocation mechanisms *vs.* revenue-raising instruments

This section identifies two water-related challenges and claims that pricing contributes to address them both. Recent trends regarding these challenges are explored in more detail in Chapter 3.

On the one hand, water is an increasingly scarce resource (see OECD, 2008 for a review of trends, at global and regional levels) and has to be allocated to where it creates more benefits for society. Experience has shown that the same benefits can be obtained using less raw water and improved technologies. For instance, under certain circumstances, it may be less costly to repair leakage than to build a new dam; or it may be preferable to abandon certain activities that extract low added value from water. Thus, water has to be managed and carefully allocated.

The water resource management challenge calls for greater care in allocating water among competing uses and protecting it against wasteful use and contamination. In this context, pricing can be an allocation mechanism, directing water where it is more valuably employed. It is also an instrument to manage demand (doing more with less water) or to enhance the productivity of existing water assets (doing more with the same amount of water), *e.g.* by providing incentives for the development and adoption of water-saving technologies. Pricing can contribute to this by informing users (including polluters) about the economic value of the resource, so that their decisions take this value into consideration. Its effectiveness depends on how well different users respond to economic incentives.[1]

On the other hand, access to sustainable and affordable water services has to be ensured to the largest possible number of people. The water supply and sanitation (WSS) challenge has to do with underinvestment, but also with wasteful or ill-conceived investment. In this context, water pricing is first and foremost a financing mechanism: it generates revenues that can be used to maintain, renew and extend the infrastructure, (when and where appropriate). The capacity to generate revenues and recover costs contributes to the financial sustainability of service provision, including by attracting financial and other resources to the sector. By encouraging efficient water use, water pricing can also reduce the need for new investment.

Failure to recognise the economic and other values of water services is one of the factors that set in motion a vicious cycle of under-funding in water-related infrastructure and management activities. If benefits are not fully recognised, funds are difficult to mobilise, resulting in lower than needed investment, inadequate maintenance of infrastructure and in the difficulty to attract good quality resources to the water sector (including human resources). This in turn results in poor management of the resource and low-quality services, which reinforces the cycle by further reducing their value in the eyes of users and governments and therefore their willingness to pay for water services and to provide public funding to the sector.

Before practices in OECD and partner countries are analysed, pricing has to be understood in the larger context of water financing. We turn to this in the next section.

Pricing as one financing instrument for the water sector

In this section, pricing is considered in the context of water financing. Water-related services generate a variety of costs which have to be recovered and financed. There are three instruments to cover these costs; pricing is one of them and will be characterised *vis-à-vis* other instruments. The discussion will point at policy dilemmas which have to be reconciled to finance water-related services in a sustainable way. Two dimensions of water pricing (the level and the structure of the tariffs) have to be combined to solve these dilemmas.

The multiple elements of water-related costs

Water management activities and WSS service provision and use come at a cost. This cost comprises a variety of elements. Ignoring some of these elements of costs will eventually result in unsustainable water resource use and water services, with consequent losses in societal welfare.

Full supply or financial costs[2] are those associated with supplying water services to users without considering either the externalities of water consumption (positive or negative) or alternate uses of water (opportunity costs). These costs consist of three elements:

- operation and maintenance costs, associated with daily running of the water supply system, such as electricity for pumping, labour and repair costs;
- capital costs, covering both capital for renewal investment of existing infrastructure and new capital investment costs;
- the cost of servicing debt.

Full economic costs are the sum of full supply costs and:

- Opportunity costs (or resource costs), which reflect the scarcity value of the resource; they refer to the cost of depriving the next possible user: if that user has a higher value for the water, then there are some opportunity costs experienced by society due to this misallocation of resources.
- Economic externalities, consisting of positive externalities (for example, groundwater recharge benefits from irrigation or water reuse) and negative externalities (typically upstream diversion of water or the release of pollutants downstream within an irrigation or urban water system).

The full costs of water use equal the sum of the full supply and economic costs, plus environmental externalities. While economic externalities cover costs to producers and consumers (upstream and downstream), environmental externalities are associated with costs to public health and ecosystems. Some practical difficulties associated with the measurement of these costs are highlighted in Box 1.1.

Administrative and governance costs needed to sustain services should be added to the above, according to Cardone and Fonseca (2003). These costs include those incurred in regulating the service, institutional capacity building, and the cost of devising and implementing the policy and enabling environment for the sector. In the same vein, Rees, Winpenny and Hall (2008) argue that this should be further extended to more systematically include the costs associated with water resources management activities needed for the stewardship of the water resource base, and therefore for the long-term sustainability of service provision.

Box 1.1. **Estimating full costs: a few practical difficulties**

There is disagreement about how to estimate "full costs" of water and sanitation services, *e.g.* whether they should include abstraction and environmental costs, whether they should include the cost of storm water disposal and flood control, whether to recover capital charges relating to "stranded assets" that no longer provide a useful service, how to allocate overhead costs between different functions and users, etc.

In addition, there are difficulties with computing the resource cost of water, due to its nature as a renewable flow and non-appropriable stock. A clear difficulty derives from the distinction between "blue water" (outflows that can be abstracted and used), and "green water" (the stock of water that is stored in the land and represents the substance of living beings). Ecosystems mostly rely on both blue and green water, and human uses mainly on blue water; it is thus not possible to simply sum their requirements. This complicates the calculation of water balances, and therefore the calculation of the scarcity value of water. The correct determination of resource cost would require an understanding of the relation that green and blue water have in a given territorial context. This relation is for the most part site-specific and depends on external circumstances, such as climate variability. Hence, it is very difficult to compute proper resource and environmental costs that can be simply added to the other economic costs.

Financing the water sector: sustainable cost recovery, revenues and repayable finance

Pricing is not the only financing instrument available to cover the costs associated with water-related services. Experience shows the difficulties of achieving full cost recovery through tariffs alone in the water sector (Box 1.2). This is not only true for cost elements going beyond supply costs, such as the cost of "institutional" components, but also for investment costs, which are often not covered through consumer tariffs. The OECD 2007-08 Survey sheds some light on cost recovery practices in OECD countries (see Chapter 3 and Tables 3.1, 3.2 and 3.3).

These difficulties are reflected in the EU Water Framework Directive, which requires member states to "take account of the principle of recovery of the costs of water services, including environmental and resource costs" (Article 9, Directive 2000/60/EC), but allows for some flexibility and lower recovery rates if appropriately justified (Article 9).

Box 1.2. **Some examples of limited cost recovery through tariffs**

Very few countries have attempted to cover full economic and environmental costs in water prices, a notable exception being Denmark. In some countries (*e.g.* Italy, United Kingdom) at least part of the "governance" costs are included in the retail tariff, but the issue is open to controversy (*e.g.* in Italy the practice has been challenged and sometimes prohibited).

In the United States, a system of federal grants underpins state revolving funds that lend at favourable rates to local authorities for investment in WSS infrastructure, and their bond issues are interest-free.

In the majority of developing countries, WSS investment costs are financed with grants and soft loans from government or official development assistance (ODA). For instance, the Senegalese water provider claims to be in "financial equilibrium" since 2003. However, practically all its investment is funded from ODA. Uganda's National Water and Sewerage Company had its government debt converted to equity to improve its balance sheet and increase its creditworthiness.

Recognising these difficulties, the policy debate has moved away from a call for full cost recovery (FCR) through tariffs and towards the concept of sustainable cost recovery (SCR) introduced in the Camdessus report (Winpenny, 2003), which identified three main features of SCR:

- an appropriate mix of tariffs, taxes and transfers to finance recurrent and capital costs, and to leverage other forms of financing;
- predictability of public subsidies to facilitate investment (planning);
- tariff policies that are affordable to all, including the poorest, while ensuring the financial sustainability of service providers.

A sustainable cost recovery strategy for the water sector aims to sustainably cover costs through a combination of three sources of revenues: tariffs (or other charges linked with water use), taxes (in the form of subsidies from national or local governments) and transfers (from international donors or local charities) – the "3Ts" defined in OECD (2009a, 2009b). Final users and local or international taxpayers are those who actually pay for water. External sources of finance that must be repaid (loans, bonds, etc.) or compensated (equity), can only bridge the gap between finance needs and available resources, particularly for investment costs that could not be covered up front through revenues alone.

Matching costs and revenues, and combining the "3Ts" are part of an exercise in strategic financial planning[3] to balance costs and revenues at national and/or at distinct sub-national levels. There are many intermediate solutions, with different compensation and cost-sharing mechanisms. Cost recovery can be alternatively intended strictly for each individual consumer,[4] for groups of consumers, or for larger territorial aggregates. It can be all internal to the water sector or involve transfers from outside the sector. Figure 1.1 depicts a continuum of public finance instruments; in the figure, "endogenous" means resources accrued from the water sector, while "exogenous" refers to resources that are not generated from the water sector.

Not achieving FCR through tariffs is a deviation from the user-pays or polluter-pays principles (see OECD, 1972, 1974). From a financial standpoint, this may be acceptable if tariffs can be appropriately combined with other reliable sources of revenues. From an economic point of view, it is also sometimes possible to justify deviations from the two principles, particularly when benefits accrue to groups of beneficiaries that go beyond their direct users. Indeed a number of costs derive from activities which do not directly use or pollute water.

Figure 1.1. **Alternative criteria for spreading the cost of water services**

ENDOGENOUS ↑

- From water users
  - On an individual basis – marginal cost pricing with no subsidy
  - Cross-subsidies from other users through the tariff structure
  - Territorial cross-subsidies from charges linked to water use (*e.g.* France)
- Cross-subsidies from fees other than water tariff and water-related charges
  - Cross-subsidies between different services (*e.g.* electricity, gas) operated by the same authority
  - Cross-subsidies among users of the same water resource, but not charged based on water use (*e.g.* downstream beneficiaries paying to reduce pollution upstream)
- General taxation and transfers
  - Direct subsidies (grants for new investment, coverage of operational deficits)
  - Indirect subsidies (*e.g.* low interest loans, under-pricing of commodities and services supplied by the public sector)
- Transferred elsewhere as an external cost
  - Transferred to other water users (*e.g.* pollution of a river used for bathing/fishing)
  - Transferred to other generations (intergenerational externalities, *e.g.* poor maintenance of assets, public debt for covering operational expenditure, permanent contamination of water tables)

↓ EXOGENOUS

But the three sources of revenues cannot be seen as perfect substitutes. The combination of the 3Ts is not neutral and there may be a number of economic and financial reasons for favouring tariff revenues over the other two "Ts".

First, the 3Ts provide different incentives to final users; only tariffs can simultaneously serve as a revenue source and as a signal regarding the value of water resource and WSS services. Second, the 3Ts provide different managerial incentives to service providers: those who have to meet their revenue requirement primarily through tariffs may have much stronger incentives to be efficient and accountable to customers. Transparency International suggests that this may help curbing corruption.[5] Third, the need to rely on tariffs may provide adverse incentives to the extension of networks to poorer areas, as the poor can be perceived by operators as loss makers (even though this is often a misperception).

Finally, investors and financiers may perceive the reliability of the three revenues streams differently, with consequences on their willingness to fund the sector. External finance will only materialise if revenue streams are sufficient and stable, and when the overall balance between risks and returns is perceived as appropriate (see OECD 2009c; OECD, forthcoming). Attracting external (generally private) finance into the water sector requires that free cash flows, *i.e.* what is left of revenues after paying for operational costs and short-term debts, are secured and set aside over time in a predictable manner.

Consequently, what matters in prices' contribution to sustainable cost recovery are: (i) their average level; (ii) the reliability of their flow and of their automatic adjustments (*e.g.* to inflation); and (iii) the flexibility in adjusting them to unforeseen circumstances (*e.g.* exogenous shocks, such as a devaluation or a surge in the price of a critical input).

A policy dilemma: trade-offs between multiple objectives

Water policies pursue multiple objectives, which can be structured around four "sustainability dimensions" (Table 1.1; for a detailed discussion, see Massarutto, 2007a).

Table 1.1. **Four policy objectives and their components**

<table>
<tr><td>Environmental sustainability

Discourage depletion of critical natural capital<ul><li>Guarantee the preservation of ecological functions of water natural capital</li><li>Minimise the use of "supply side" solutions to water scarcity</li><li>Use efficiency
- Encourage water saving
- Discourage wasteful water use</li><li>Minimise the alteration of natural flow patterns</li></ul></td><td>Financial sustainability

Guarantee long-term reproduction of physical assets<ul><li>Compensate the resources that are used as inputs in water-related activities</li><li>Cash flow should guarantee the conservation of value of physical assets</li><li>Cost efficiency: minimise lifecycle costs of services, i.e. the creation of physical capital and operation and maintenance costs</li><li>Cost recovery should be for efficient costs only</li></ul></td></tr>
<tr><td>Economic efficiency

Water is allocated to the most beneficial uses and economic resources are not wasted<ul><li>Allocation efficiency:
- Allocate water with priority to uses with highest value to society as a whole
- Compare costs of water management and water-related services with their value, i.e. do not misallocate economic resources</li><li>Regulation should ensure optimal risk allocation among stakeholders (including users and taxpayers)</li></ul></td><td>Social concerns

Adequate access to affordable water at fair and equitable conditions<ul><li>Identify "water needs" and allocate water in a way that is not skewed by concentration of power</li><li>Structure tariffs so that lower-income users can have access to and afford to use WSS services</li><li>Achieve an equitable way to share the cost of managing water resources</li></ul></td></tr>
</table>

Environmental, economic, financial and social objectives can support one another, but sometimes they can also give rise to potential conflicts. Given the multidimensional nature of water values, governance issues may arise from mismatches between available and demanded ecological functions of the water resource and other aspects of four sustainability dimensions (see Figure 1.2).

An example of such a trade-off would be social demands on the water resource and environmental sustainability. Available ecological functions of water resources are constrained by hydrologic, technological and economic factors. To a certain extent, their availability can be incremented by adding manmade capital, but this is costly and it takes time for such interventions to balance demands. A mismatch of objectives can arise: (i) if the community expressing the demand is too poor to afford covering these costs; (ii) if costs are higher than the value of the required environmental functions (reducing economic efficiency); or (iii) if other communities/user groups are unwilling to provide support through higher taxes or cross-subsidies from higher water prices.

Figure 1.2. **Policy objectives and trade-offs that affect pricing levels and structures**

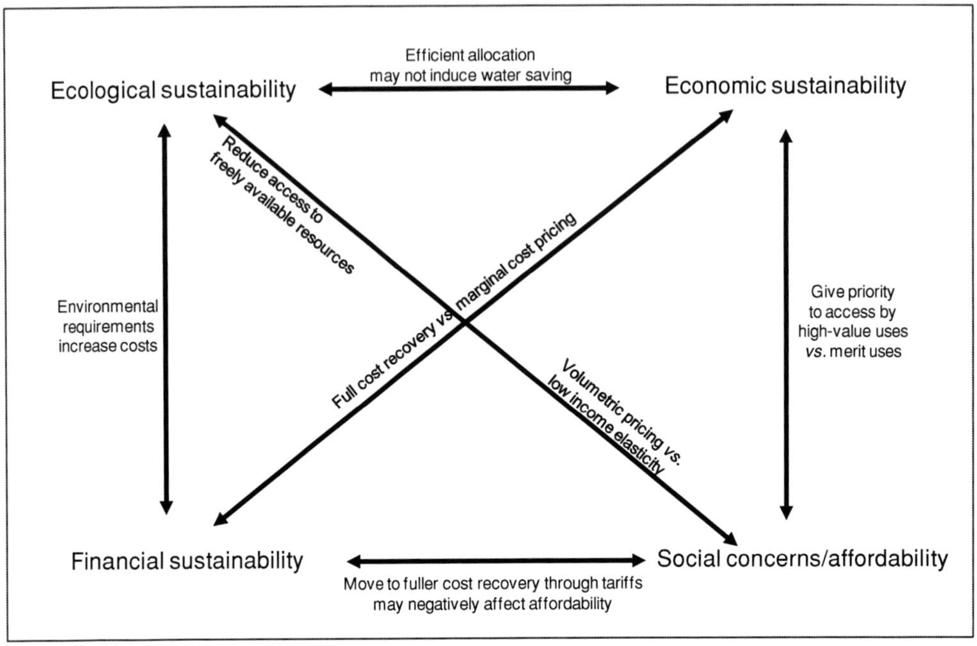

Source: Based on Massarutto, A. (2007b), "Abstraction Charges: How Can the Theory Guide Us?", presentation made at the OECD Expert Meeting "Sustainable Financing for Affordable Water Services: From Theory to Practice", November.

Trade-offs and the capacity to address them evolve over time. Income improvements may enable a low-income community to face the prices needed to obtain services that were previously unaffordable; technological improvements might reduce costs; more effective institutions might emerge; social learning processes might enable the community to accept previously unacceptable solutions (*e.g.* pricing). It follows that pricing strategies would benefit from recurrent assessments and revisions.

A synthetic discussion of selected trade-offs follows. The capacity to address these potential dilemmas relies in part in appropriate governance structures and procedures.

Financial sustainability vs. *economic efficiency*

The optimal tariff-setting rule from the point of view of economic efficiency may be inconsistent with financial sustainability and may provide adverse incentives for investment. Economic efficiency objectives focus on allocating water resources to the most beneficial uses for society, avoiding over-investment, using existing facilities efficiently, and ensuring the operational efficiency of water systems. The economic literature on water pricing generally recognises long-term marginal cost (MC) pricing as the first-best pricing option.

However, from a utility perspective, MC pricing is inconsistent with the need to ensure stable revenue flows that ensure the accumulation of funds for future investment. This, and other practical difficulties, has led to the widespread use of average incremental cost pricing. The general consensus is that fixed costs of infrastructure should be paid by a separate lump sum. In economic terms, it is irrelevant whether the lump sum originates from the general budget or in other ways (*e.g.* earmarked taxes, connection charges, individualised fees), as none provide incentives to reduce consumption. However, each alternative has distinctive distributive outcomes.

Affordability: financial sustainability of water operators vs. *social considerations*

The issue that most often sparks controversy in the debate regarding WSS tariffs is the potential conflict between financial sustainability through increased tariffs and access to affordable services. In reality, the controversy may rest on a number of misunderstandings, and favouring one objective at the expense of the other may undermine the possibility of achieving either of them.

Typically, keeping tariffs artificially low for all customers, including those who can afford the full price of the service, may lead to the vicious cycle of collapsing infrastructure and deteriorating services, which tends to hurt the poor the most. In situations where service coverage is incomplete, lack of funding prevents the extension of services to unserved households - generally the poorest – forcing them to pay much more to obtain services that can be of inferior quality. Even where the poor are connected, deteriorating services may hurt them disproportionately, as alternative options (*e.g.* bottled water) may be expensive.

Affordability can be assessed by comparing the price of water services (the water bill) with the capacity-to-pay of final users. This capacity can be measured using a variety of indicators, including disposable income, household expenditures, or expenditure on other essential services (*e.g.* energy). Affordability can be assessed at a macro level (all households in an area or country), or at a micro level (specific groups, such as lower income or other vulnerable people).

There is no absolute level for affordability. International thresholds exist and can be a useful basis for preliminary assessments at national level: donors, international financial institutions and other international organisations often quote the figures of 3-5% of disposable income or household expenditure. But there is no official statement or recommendation: thresholds need to be discussed and determined locally, and may need to be revised over time.

Macro affordability issues can be addressed in the context of strategic financial planning (see OECD, 2009a for a discussion and illustrations). Planning allows to minimise costs (by revisiting the level of service, the technical options, or the sequence of projects) and increase revenues (through tariffs, taxes, or transfers).

Micro affordability indicates that tariffs are too burdensome for specific population groups. What matters then is the way costs are allocated across different groups (including other domestic users) and the choice of cross-subsidisation mechanisms. From the perspective of tariff design, the protection of vulnerable groups is less concerned with average tariff levels and more with the definition of a tariff structure that includes redistributive criteria, or non-tariff instruments (*e.g.* income support, facilitating payments) to carefully target vulnerable groups.

The effectiveness of any tariff structure is reduced if poor households do not have access to the service whose consumption is subsidised. If access is systematically skewed against the poor, any tariff structure that provides negative incentives (or insufficient funds) for the extension of services will be regressive. In these cases, subsidising access has been demonstrated to be more effective than subsidising consumption (Komives *et al.*, 2005).

The introduction of social tariffs needs not be at odds with financial sustainability objectives. Social tariffs can be compensated by cross-subsidisation mechanisms (*e.g.* from other, better-off user groups), and by additional revenues from taxes or transfers. Pricing strategies may sometimes be able to balance affordability and financial sustainability objectives by themselves. However, they will more often need to be combined with other policy instruments.

From water pricing to tariffs

The previous discussion points at tariff structure as an essential dimension of water prices. It follows that water pricing policies deal at the same time with tariff levels and structures.

Revenues from water prices derive from the following components:

- A one-time connection fee, to gain access to the service.
- A recurrent fixed charge (sometimes known as a standing charge or flat fee) that can be uniform across customers or linked to some customer characteristic (*e.g.* size of supply pipe or meter flow capacity; property value; number of water-using appliances).
- If a metering system is in place, a volumetric rate, which, when multiplied by the volume of water consumed in a charging period, gives rise to the volumetric charge for that period.
- In some circumstances, a minimum charge is paid for each period, regardless of consumption.

Different forms and combinations of the elements above (with or without a connection charge) give rise to the following tariff structures:

- Flat rates (either uniform or differentiated): in a non-metered environment, customers pay a flat rate regardless of their consumption. This can be uniform, or differentiated based on customer characteristics, season, etc.
- Single volumetric rates with/without uniform or differentiated fixed charges: in a metered environment, a single rate per cubic metre is applied regardless of volume consumed. This can be charged with or without a recurrent fixed charge. The fixed charge can also be negative (a coupon). Fixed charges and coupons can be uniform or vary according to customer characteristics.
- Increasing block tariffs (IBTs): the volumetric charge changes in steps with volumes consumed.
- Adjusted IBTs: either the volumetric rates applied to each block or the size of the blocks are adjusted based on specific customer characteristics (*e.g.* family size, income).
- Decreasing block tariffs: the volumetric rates decline with successive consumption blocks.

When tariffs apply to volume, it is noteworthy that revenues depend both on the tariff and on the volume which is billed. Additional revenues can be raised with the same tariff, by decreasing unaccounted-for water (*e.g.* by reducing leakages).

Policy issues

Two policy questions derive from the discussion which cut across the OECD Horizontal Water Programme:

- What proportion of costs is recovered by revenues from tariffs? What alternative cost-sharing mechanisms may be appropriate for different water-related activities?
- Are average tariff levels affordable for all? And if not, are adequate tariff structures, including social tariffs, or other income-support mechanisms being adopted?

The OECD 2007-08 Survey was designed to address these questions. The data collected on water prices and tariffs are presented in the next chapter. The policy issues are analysed in Chapter 3.

Notes

1. OECD (2010) indicates, for instance, that agriculture water users respond imperfectly to water prices, with threshold effects and other peculiarities that require a careful analysis of the full set of incentives that these users face before introducing a pricing strategy,
2. The classification up to the definition of full costs follows Rogers, Bhatia and Huber (1998), although slightly different versions exist in the literature.
3. See OECD (2009a) for a full-fledged discussion of the strategic financial planning process.
4. The extreme case being price discrimination to reflect the different costs of serving each customer.
5. Transparency International (2008) provides estimates of the cost of corruption in the water sector. While the exact figures can be debated, the orders of magnitude are significant.

References

Cardone, R. and C. Fonseca (2003), "Financing and Cost Recovery", Thematic Overview Paper, No. 7, IRC.

Komives, K., *et al.* (2005), *Water, Electricity, and the Poor — Who Benefits from Utility Subsidies?*, The World Bank, Washington, DC.

Massarutto, A. (2007a), "Water Pricing and Full Cost Recovery of Water Services: Economic Incentive or Instrument of Public Finance?", *Water Policy*, Vol. 9, pp. 591-613.

Massarutto, A. (2007b), "Abstraction Charges: How Can the Theory Guide Us?", presentation made at the OECD Expert Meeting "Sustainable Financing for Affordable Water Services: From Theory to Practice", November.

OECD (1972), "Recommendation on Guiding Principles Concerning International Economic Aspects of Environmental Policies", C(72)178.

OECD (1974), "Recommendation on the Implementation of the Polluter-Pays Principle", C(74)223.

OECD (2008), *OECD Environmental Outlook to 2030*, OECD Publishing, DOI: *http://dx.doi.org/10.1787/9789264040519-en*.

OECD (2009a), "Strategic Financial Planning for Water Supply and Sanitation", OECD internal document, *www.oecd.org/water*.

OECD (2009b), *Managing Water for All: An OECD Perspective on Pricing and Financing*, OECD Publishing, DOI: *http://dx.doi.org/10.1787/9789264059498-en*.

OECD (2009c), *Private Sector Participation in Water Infrastructure: OECD Checklist for Public Action*, OECD Publishing, DOI: *http://dx.doi.org/10.1787/9789264059221-en*.

OECD (2010), *Sustainable Management of Water Resources in Agriculture*, OECD Publishing, *www.oecd.org/tad/env*.

OECD (forthcoming), *Innovative Financing Mechanisms for the Water Sector: An OECD Perspective,* OECD Publishing.

Rees, C., J. Winpenny and A. Hall (2008), "Water Financing and Governance", *GWP TEC Background Paper* 12.

Rogers, P., R. Bhatia and A. Huber (1998), "Water as a Social and Economic Good: How to Put the Principle into Practice", *TAC Background Papers,* No.2, Global Water Partnership, Stockholm, Sweden, *www.gwpforum.org/gwp/library/TAC2.PDF.*

Transparency International (2008), "Global Corruption Report", Transparency International, Berlin.

Winpenny, J. (2003), "Financing Water for All", report of the World Panel on Financing Water Infrastructure, chaired by Michel Camdessus, *www.financingwaterforall.org.*

Chapter 2

Water pricing in OECD countries: state of play

This chapter provides a set of robust and comparable data on water prices in selected OECD countries. It covers abstraction charges, pollution charges, pricing water supply and sanitation for households and industrial users (including connection charges).

The data show that, generally speaking, efforts are being made in OECD countries to better reflect the costs and externalities related to water use by households and industrial users. This is reflected in the deployment of abstraction and pollution charges, in the level of prices (which have increased, at times substantially, over the last decade) and in the structure of tariffs (which better reflect consumption and treatment costs).

In 1999, the OECD carried out an extensive review of water pricing practices in OECD countries (see OECD, 1999; partly updated in OECD, 2003a). In 2007-08, the review of OECD country experiences was updated. The analysis was extended to include experiences in non-OECD countries. The OECD 2007-08 Survey covered water pricing for households and industrial users. It did not cover pricing of water in agriculture, as this was the object of another component (OECD, 2010) of the OECD Horizontal Water Programme, of which this work is part.

A note on method

The method used for the survey and for analysing data recognises that: (i) pricing, cost and other relevant data on water-related services is fundamentally local, that any aggregation or averaging exercise implies a loss of information; (ii) choices regarding sampling and aggregation affect national values; and (iii) extreme care should therefore be taken in proposing cross-country comparison on such variables.

The survey builds on a variety of data.

Information on abstraction and pollution charges was extracted from the OECD/European Environment Agency (EEA) database on environmental taxes and fees.

For water supply and sanitation pricing levels and structures in OECD countries, a first round of collection and analysis was carried out based on publicly available data. Information gaps were identified and a questionnaire was sent to member countries. Country experts were given two options: they could compile data at the national level, documenting the methodology used for their aggregation, or provide disaggregated data collected from local service providers (or a sample thereof, to be defined by the national experts). In addition to tariff levels and structures, questions were asked on the cost recovery levels.

In parallel, two surveys on pricing levels and structures were carried out by Global Water Intelligence (GWI) in 2007 and 2008. These covered over 150 cities in all OECD countries, and 100 cities in non-OECD countries on all continents, including key emerging economies (Brazil, China, India, Indonesia, South Africa), Eastern Europe, the Caucasus and Central Asia (EECCA). While GWI has been carrying out surveys since 2003, the OECD requested additional information and an extension of the sample of providers/cities included in the survey.

Additional empirical evidence came from the data collected biennially by the International Water Association (IWA) and the IB-Net database managed by the World Bank.

To complement databases, five sets of case studies were used, building on work done in the context of the OECD Horizontal Water Programme and of the Environmental Action Plan (EAP Task Force). Case studies were meant to assess: (i) recent experiences with tariff structures and instruments aimed to improve the capacity to deal with affordability issues; and (ii) the difficulties of comparing cost recovery levels and financial sustainability of service providers.

The different sources are not readily compatible, as scopes, definitions and methodologies differ; a comprehensive set of data collected from the OECD and GWI surveys is appended. It was decided to rely as much as possible on data endorsed by the countries (for OECD members) and to avoid aggregation of local data into national indicators.

Pricing water resource management

A range of mechanisms can be used to transfer some of the costs of water resources management activities to their beneficiaries, including (see Rees, Winpenny and Hall, 2008, for more information):

- Regulatory levies: These are increasingly being used to recover regulatory costs from the regulated parties. In a number of countries, the regulation of water abstraction and wastewater discharges is funded through license fees and charges.
- Pollution and abstraction charges or taxes: They are based on the user-pays and polluter-pays principles. They include charges associated with non-tradable abstraction, consumption or pollution permits, and effluent or pollution charges. They aim to recover costs and to internalise negative externalities associated with water abstractions or polluting activities. As a proxy, most charges are set to cover the costs of investment programmes aimed at environmental improvements.
- Payments for ecosystem services: In some cases, downstream beneficiaries pay to regulate or preserve or restore upstream environments (*e.g.* flood management), as they benefit from activities made by others to reduce water consumption or pollution. Upstream land and water users/polluters receive compensation to provide environmental services and avoid damaging practices.

- Permit markets: Markets for abstraction and pollution permits are created to: (i) facilitate market-regulated water permit reallocation under scarcity conditions; and (ii) provide incentives for pollution abatement and technological improvements, by allowing polluters who can outperform environmental standards to sell excess pollution rights.

This section focuses on abstraction charges and pollution charges. It provides a brief overview of their use in OECD countries and selected non-OECD countries.

Abstraction charges in OECD countries

Since 1998, the OECD has maintained a database on environmentally related taxes, which is regularly updated by experts of ministries of finance and ministries of environment in member countries.[1]

The form taken by taxes and fees/charges and the basis for their calculation vary considerably by country and sector. Charges can take the form of a nominal license fee linked to an abstraction permit regime, a volumetric charge varying with abstraction or consumption volumes, or a flat or variable charge linked with other criteria (*e.g.* area of industrial estates).

OECD countries are gaining experience with abstraction charges, pollution/effluent charges and other economic instruments – such as tradable water use permits – to achieve more economically efficient and environmentally sustainable abstraction and allocation among competing uses.

Tables 2.1 and 2.2 combine the most recent information in the OECD/EEA database and the responses by member countries to the OECD 2007-08 Survey. They may therefore not be comprehensive. The third column of Table 2.2 indicates whether the abstraction charges are paid by suppliers of retail water supply and sanitation (WSS) services, and passed on to their customers through retail tariffs.[2] In both tables, "source" indicates surface *vs.* groundwater.

In most countries, abstraction charges are designed with the objective of providing funding for water resources management or for watershed protection activities. Despite this, Table 2.1 shows that abstraction charges tend to be relatively low, and value ranges are quite broad. In the limited sample available, higher charges are imposed on groundwater than on surface water.

In Australia's New South Wales region, the Independent Pricing and Regulatory Tribunal sets Bulk Water Prices that the State Water Corporation and Department of Water and Energy (which in 2007 took over the responsibilities of the Department of Natural Resources) can charge to irrigators, industrial users and providers of drinking water. A key preoccupation is full cost recovery ("cost-reflective pricing"), which has been achieved in most regulated rivers. A two-part tariff was chosen, which includes a fixed charge and a volumetric charge that is the sum of an entitlement charge (on the water abstraction rights) and a usage charge (on water actually abstracted). Unregulated rivers and groundwater are still far from cost recovery levels, but the 2006 determination of bulk prices for 2006-10 increased fees paid by irrigators, while allowing them to move to a two-part tariff if they are willing to be metered. As a result of the full cost recovery policy, charges have risen by inflation plus 15% per annum between 2001 and 2005 for regulated rivers and inflation plus 20% for unregulated rivers. Proposals for prices are subject to public debate, including public meetings (NSW Independent Pricing and Regulatory Tribunal, 2004).

Table 2.1. **Abstraction charges by sector and origin in selected OECD countries, 2008**

USD per cubic metre

Country	Source	Drinking water	Agriculture	Industry	Other
Belgium					
Flanders	Groundwater	0.08	0.08	0.08	0.08
Wallonia	Surface or groundwater[1]	0.11			
	Groundwater		0.04	0.04	0.04
Czech Republic	Surface water			0.03	0.11
	Groundwater	0.12	..	..	0.18
Germany[2]	Surface water	0.03	0.01	0.01	0.01
	Groundwater	0.14	..	0.05	0.06
Hungary[3]	Surface water	0.003	..	0.003	..
	Groundwater	0.003	..	0.003	..
Netherlands	Surface water				
	Groundwater	0.27	..	..	..
Poland[4]	Surface water	0.01	..	..	0.02
	Groundwater	0.02	..	0.03	0.04

Notes:

1. Charges for non-drinking water abstraction: it is not specified which sectors are charged.
2. State/regional abstraction charge of USD 0.19/m3 (average value) not included.
3. Average values, charges vary on the basis of water sources and region.
4. Charges vary on the basis of water quality and region.

Source: OECD/EEA database on instruments used for environmental policy and natural resources management.

Table 2.2. **Features of abstraction charges and fees in OECD countries**

	Charges for direct water abstraction	Charges included in retail WSS tariffs	Proceedings collected by	Basis for charging	Differentiated user type	Charges differentiated by other characteristics
Austria[1]	..	..	..	..	..	..
Belgium						
Flanders	Yes	Yes	Regional administrations (different for groundwater and surface water)	Capacity and Actual use	Yes	Source, discount for return flows, water scarcity
Brussels	No	No	..	..	..	..
Wallonia	Yes	Yes	Regional administration	Actual use	No	Volume and water scarcity
Canada	Yes	Yes (varies by municipalities)	Varies by municipalities	Capacity	Yes	Volume and province
Czech Republic	No	No	..	Actual use	No	Source, river basin
Denmark	Yes	Yes	Municipalities, from 2009 tax authorities	Capacity	No	..
Finland	No	No	..	..	..	..
France	Yes	Yes	River basin agencies	Capacity	Yes	Source, location and river basin agency
Germany	Yes	Yes	Regional administrations	..	Yes	Source and Länders
Hungary	Yes	..	..	..	Yes	Volume
Italy	Yes	Yes (regional basis)	Regional authorities	..	Yes	Location
Japan	Yes	..	Local authorities	..	..	Source and location
Korea	Yes	No	Municipalities	Actual use (partially)	No	..
Mexico	Yes	Yes	Treasury of the Federal Government	Capacity and Actual use	Yes	Source and location
Netherlands	Yes (Groundwater)	..	..	Actual use	No	..
Poland	Yes	..	..	..	Yes	Source and location
Portugal	Yes	Yes	Water Protection Fund, River Basin Authority, INAG (national regulator)	Actual use	No	..
Slovak Republic	Yes	..	..	..	..	..
Spain	No	Yes	River basin, local administrations	Actual use	Yes	River basin
Sweden	Yes	No	..	Actual use	No	..
Switzerland	Yes	Yes	Cantonal authority	Capacity	No	..
United Kingdom						
N. Ireland[2]	Yes	No	Government agency	Capacity	Yes	..

1. The latest *OECD Environmental Performance Review* (OECD, 2003b) reports "water use fees" but the OECD/EEA database has no record.

2. Country reply to OECD 2007-08 Survey.

Source: OECD/EEA database on instruments used for environmental policy and natural resources management, national reports.

In Germany, abstraction charges have been introduced with the dual objective of decreasing abstractions and raising revenue for use in environmental protection measures. Of the 16 Federal States, 11 have water abstraction fees in place (see Gaulke, 2007). The fees range from EUR 0.015/m^3 in Saxony (EUR 3.55 million in 2006) to EUR 0.31/m^3 in Berlin (EUR 57.6 million in 2006) (Ecologic, 2008). Revenues have been used for nature conservation, protection of ground and surface water, reforestation, soil protection and decontamination. In seven of these Federal States (in Berlin, for example), (part of) the money is earmarked for groundwater protection.

In Belgium's Flanders region, incentives to use water resources efficiently are provided through groundwater charges whose price per cubic metre increases with total amount of groundwater pumped. Charges also vary according to the aquifer in which the extraction well is situated and the existing pressure head of the groundwater. Drinking water companies pay a higher charge. A decreasing block system is used to charge for surface water abstraction, whose charges are lower than for groundwater.

In Portugal, until recently charges existed in law but were not collected, so water resources management relied primarily on command and control measures. Starting in mid-2008, water supply and sanitation service providers are obliged to include abstraction charges in the retail tariffs dependent on the actual use and the type of user. The proceedings are collected for 50% in a "water protection fund", for 40% by the Basin Water Authorities (ABH), and for 10% by the National Water Authority in charge of water resources management (INAG).[3]

In Canada, most provinces levy licence fees to major water users for access to the resource. The provincial licence fees for water are related to the cost of administering the licensing programme. These are regulatory levies and not abstraction charges.

In most cases, charges are collected and retained locally. Only in a few cases does the revenue merge into general taxation. This is the case in Denmark and Mexico, while in Germany some abstraction charges go into the budgets of some of the *Länder*.

Pollution charges in OECD countries

Table 2.3 summarises the characteristics of pollution charges in selected countries for which information was either received from national experts or available in the OECD/EEA database (unless otherwise specified). These charges can represent a significant share of the water bill (about one-third of WSS bills for households, in the case of France, as an illustration).

Table 2.3. **Features of pollution charges in selected OECD countries**

	In place	Levied by	Based on	Description (OECD/EEA database)	Charges included in the water tariff	Fines
Australia[1]	Yes	State/Territory Jurisdiction	Volume and pollution content, location	Charges levied on 17 types of pollutants	..	Yes
Austria[1]	Yes	Municipalities	Pollution load	..	..	Yes
Belgium[1]	Yes	Local authorities/Fund	Pollution unit/pollution content	Regional manure charges by volume; wastewater charges for households and industry	..	..
Flanders	Yes	Regional authorities	Volume and pollution content	..	..	..
Brussels	Yes	Regional authorities	Volume and pollution content	..	..	..
Wallonia	Yes	Regional authorities	Volume and pollution content	..	..	..
Canada[1]	Yes	Provinces/ Municipalities	Pollution content	Charges for industrial effluents by quantity and province	Yes	..
Czech Republic[1]	Yes	State Environmental Fund	Pollution content	Waste water effluents charges by quantity	..	Yes
Denmark	Yes	Municipalities	Pollution content/water consumption	Taxes on waste water. Agriculture is exempt	Yes	Yes
Finland	No	..	..	No charge	..	Yes
France	Yes	River basin agencies	Pollution content/users	Charge per pollutant varies according to user and water river basin agency	Yes	Yes
Germany[2]	Yes	Länders	Pollution unit/pollution content	Based on concentration and noxiousness of specific pollutants	..	Yes
Greece	No	..	..	..	..	
Hungary	Yes		Pollution content	Charges for different pollutants by vulnerability of recipient waters	..	Yes
Iceland	No	..	..	..	..	..
Ireland	No	..	..	..	..	..

Table 2.3. **Features of pollution charges in selected OECD countries** *(continued)*

	In place	Levied by	Based on	Description (OECD/EEA database)	Charges included in the water tariff	Fines
Italy	Yes	..	..	Wastewater charge by type of use	..	
Japan	..	..	..	Wastewater charge by type of use	..	..
Korea[1]	Yes	..	Volume and type of pollutant	18 types of pollutants by user and quantity	Yes (partially)	..
Luxembourg	No	..	..	..	..	..
Mexico	Yes	..	Receiving body, location, volume and pollution content	Receiving body, location, volume and pollution content	No	..
Netherlands[1]	Yes	Central gov. (V&W)/water boards	Pollution load Large industrials users are closely monitored	BOD, COD and heavy metals, by province, for large polluters		Yes (for large polluter)
New Zealand	No	..	..	..	..	..
Norway	Yes	Municipalities	Pollution content	Pesticides	..	..
Poland	Yes	Voivodships/ Regional Boards for Water Management	Volume and type of pollutant	Based on type of pollutant, industrial sectors, and receiving bodies	..	Yes
Portugal	Yes	..	Pollution content and toxicity	Effluent charges for industry	No	..
Slovak Republic	Yes	State Environmental Fund	Pollution content	Based on type of pollutants	..	Yes
Spain	Yes	River Basin Authority	Pollution content and location	Pollution parameters and location	Yes	..
Sweden	Yes	Municipalities	Pollution content	..	No	Yes (oil spills)
Switzerland	No	..	..	..	..	..

Table 2.3. **Features of pollution charges in selected OECD countries** *(continued)*

	In place	Levied by	Based on	Description (OECD/EEA database)	Charges included in the water tariff	Fines
United Kingdom	..	..	..	..	..	..
England and Wales	Yes	..	..	Environmental impact of effluent volume and toxicity according to the Mogden formula	..	..
N. Ireland	..	..	..	..	No	..
United States	Yes	Water networks	Pollution content or volume	..	..	..

Notes:

1. OECD (2006), *OECD Environmental Performance Reviews: Korea 2006*, OECD Publishing, DOI: *http://dx.doi.org/10.1787/9789264024045-en.*

2. BGW (German Water Association) (2008), "Profile of the German Water Industry", *www.dvgw.de/fileadmin/dvgw/wasser/organisation/branchenbild2008_en.pdf.*

3. Ecologic (2008), "Water Pricing and Costs of Services in Germany".

4. OECD (2008a), *OECD Environmental Performance Reviews: Hungary 2008*, OECD Publishing, DOI: *http://dx.doi.org/10.1787/9789264049284-en.*

Source: National reports prepared for the OECD 2007-08 Survey and OECD/EEA database on instruments used for environmental policy and natural resources management (unless otherwise specified).

More countries reported the use of pollution charges than abstraction charges. Pollution charges can be linked to different characteristics of the polluter (*e.g.* its sector, processes), the effluents (volume or pollutant concentration) or the recipient water body. In Belgium, pollution fees are based on measured concentration or conversion coefficients for organic matter and suspended solids, heavy metals, nutrients (N and P) and cooling water. In Mexico charges are linked with discharges' quantity and strength (in excess of permissible biochemical oxygen demand [BOD], chemical oxygen demand [COD] and total suspended solids [TSS] levels) and with the carrying capacity of recipient bodies, which is differentiated for industrial and municipal wastewater discharges. In Hungary, the "environmental load charge" applies to all activities that require a discharge permit (*e.g.* wastewater companies) and takes into account both the vulnerability of receiving waters and the technologies used for wastewater treatment and sludge disposal; to strengthen the incentive nature of the measure, a rebate applies for polluters that adopt approved pollution reduction measures (pre-treatment) and fines apply for discharges in excess of permits (OECD, 2008a).

In most cases, fees/charges are collected at the local level – but only seldom at the river basin level – to finance environmental activities. In some instances (*e.g.* Czech Republic and Slovak Republic), revenues are collected nationally but allocated to specific national environmental funds. In the Netherlands, charges apply to industrial and municipal discharges to state waters; they are linked with pollution loads which are computed based on coefficients and converted into population equivalent (p.e.), but are directly measured for large polluters; charges vary according to provinces to reflect pollution abatement costs; revenue goes to the central government but is earmarked to finance water and wastewater management activities.

In Australia, the States/Territories of New South Wales, Victoria and South Australia are operating pollution charge systems. These systems were initially set up to recover the administrative costs of licensing, monitoring and enforcement, but in recent years, incentives for license holders to continuously reduce their discharges to water have become prominent. In New South Wales, a load-based licensing (LBL) scheme was introduced in 1999 to link license fees to pollutant emissions and to provide incentives to drive down pollution. In Victoria, the fee structure for pollution licenses under the Environment Protection (Fees) Regulations Act 2001 is also designed to provide incentives for license holders to reduce their discharges and emissions (OECD, 2008b).

In Denmark, a wastewater tax is imposed on water consumers, on top of tariffs for wastewater treatment services. It applies to both households and industry, but not to agriculture. The tax is due for the total discharge of nitrogen, phosphorus and organic matter (BOD), and is additional to any charge paid to obtain a license for a wastewater treatment plant. Single (scattered) households in the countryside must also pay the tax, based on measured water consumption and standard treatment rates dependent on local treatment efficiency. There is a uniform rate nationwide whatever the quality objective or diluting capacity of receiving waters (OECD, 2008c).

In some instances, countries adopted alternative cost allocation mechanisms that recognise the existence of a broader set of beneficiaries. Under the principle of compensation recognised under Korean law, revenues collected from downstream beneficiaries are used to compensate upstream residents for losses due to land use regulation, an important step towards truly integrated water and land management at the river basin level. In the EU and the United States, farmers are paid for a variety of environmental stewardship measures, including reducing nitrate contamination. While these have been primarily funded through public budget allocation - implicitly recognising society as a whole as a beneficiary – it is possible to design alternative funding mechanisms with specific fees or levies on water bills, fees for recreational uses or fisheries and levies on flood plain dwellers, so as to allocate costs more directly to direct beneficiaries.

Pricing water supply and sanitation for households

This section reports the main data collected on prices for water supply and sanitation services provided to households in OECD countries. It covers price levels and tariff structures.

Tariff levels for water supply and sanitation

Figure 2.1 shows the average price per cubic metre of WSS services faced by a household in different OECD countries. In an attempt to improve comparability, the OECD 2007-08 Survey requested country experts to provide information on the average bill that a representative household consuming 15 m^3 per month would pay. Twelve countries/regions provided this information. For the other ten included in the figure, data from publicly available sources were used and validated by country experts. In both cases, average tariffs include the relevant volumetric charge, recurrent fixed charges (but not connection or other one-time charges) transformed into a volumetric-equivalent rate, and any indirect taxes levied on the bill.

Figure 2.1. **Unit price of water supply and sanitation services to households, including taxes, in OECD countries, 2008**

USD per cubic metre

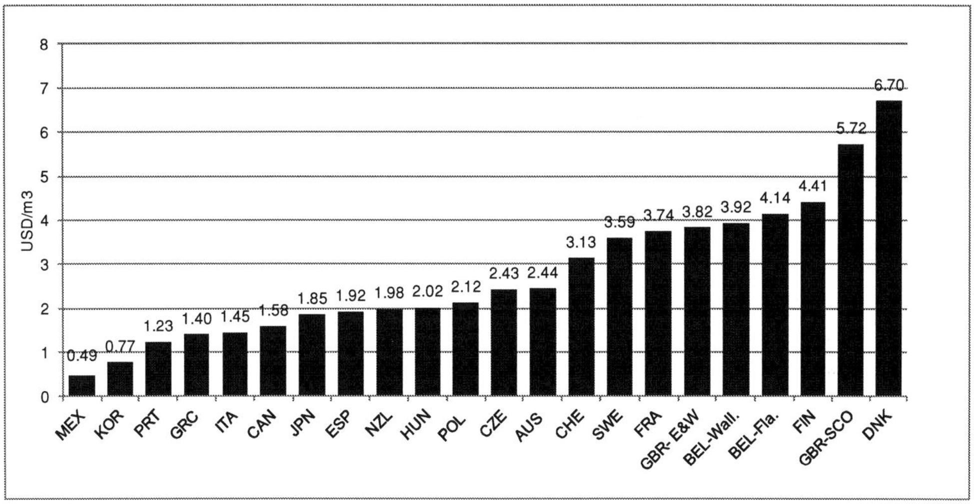

Source: OECD estimates based on country replies to the OECD 2007-08 Survey when available, or public sources validated by the countries.

The data shows large discrepancies across countries: prices can vary from a factor 10 or more. This derives from differences in the cost of provision of the service. This reflects policy choices as well. According to the data, Denmark has distinctively high prices for water supply and sanitation, which reflect efforts to incorporate as much of the economic and other costs of the services into tariffs. The apparently high unit price for Scotland may result from calculations, as it is computed from an estimated average bill for non-metered households (see the methodological information in Table A.3 in Annex A). Two countries (Mexico and Korea) have tariffs below USD 1/m^3. Prices in the other 18 countries that responded to the OECD 2007-08 Survey range between USD 1.23 and USD 4.41 per cubic metre.

The GWI survey can be used to account for the variation of prices in one single country, confirming that national averages should be interpreted with caution (see Figure 2.2). It is noteworthy that GWI collects data from operators who do not consider some levies as part of the tariff when they are separately charged by other entities; this was typically the case for Dutch cities where GWI data do not cover wastewater levies charged by regional authorities.

Figure 2.2. **Variation of local prices for water supply and sanitation in OECD countries, 2008**

USD per cubic metre

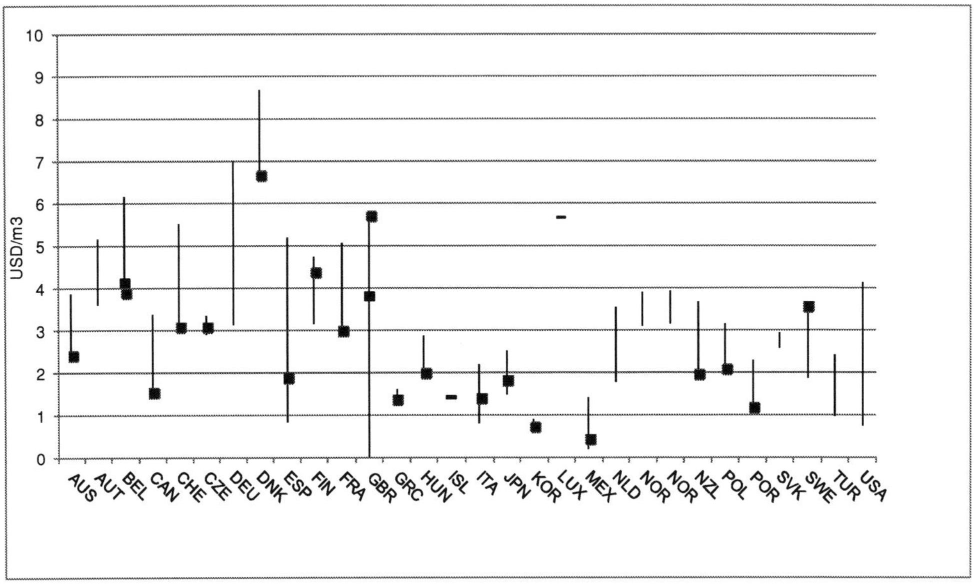

Notes:

1. Squares indicate OECD estimates based on country replies to the OECD 2007-08 Survey when available, or public sources validated by the countries.

2. The vertical lines reflect data on selected municipalities compiled by GWI.

3. The mark – for ISL and LUX indicates that data refer to one municipality only.

4. BEL: OECD data refer to the Flemish and Walloon regions.

5. GBR: OECD data refer to England and Wales and Scotland.

Source: OECD 2007-08 Survey and GWI survey.

Very significant variations exist across non-OECD countries as well (see Figure 2.3). National averages for non-OECD countries are computed on the basis of the GWI survey. Tariff levels in non-OECD countries tend to be much lower than in the OECD. Even in the water-scarce Middle-East/North Africa region, 6 out of the 13 countries in the sample had tariffs below USD 0.15/m3, providing little incentive for water use efficiency and contributing little to cost recovery. In EECCA and Sub-Saharan Africa, no tariffs in the sample reach the USD 1 mark, raising doubts on their capacity to significantly contribute to cost recovery. However, some countries in Asia (Singapore), Latin America (Bahamas, Brazil, Chile) and the Middle-East/North Africa region (Israel, Oman, Palestine, Qatar, United Arab Emirates [UAE]) have tariffs above USD 1/m3.

Figure 2.3. **Variation of local average tariffs for water supply and sanitation in non-OECD countries, by region, 2008**

USD per cubic metre

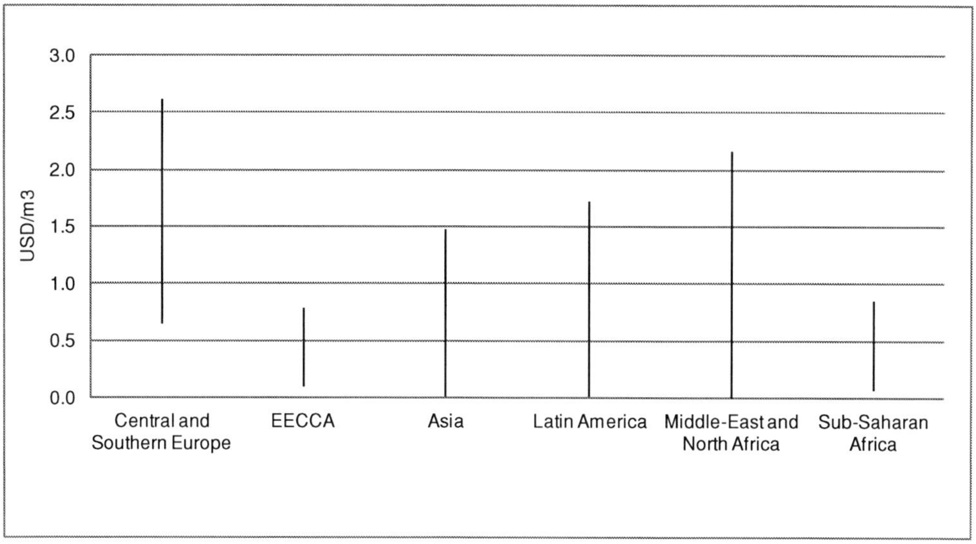

Notes: The vertical lines reflect data on selected municipalities compiled by GWI.

Source: OECD elaboration on GWI 2008 database; GWI computes national average tariffs from local tariffs at utility level.

It follows that the analysis of tariff levels needs to take place at the local level and that national and international benchmarking on tariff levels should be carried out very carefully.

Figure 2.4 presents separate average tariffs for water and for wastewater services in USD. The sample differs slightly from the previous figure, as separate data are not always available. The data show that in half of the countries, wastewater services can represent a higher share of the water bill than water supply.

Figure 2.4. **Comparison of unit prices of water services and wastewater services to households, including taxes, in OECD countries, 2008**

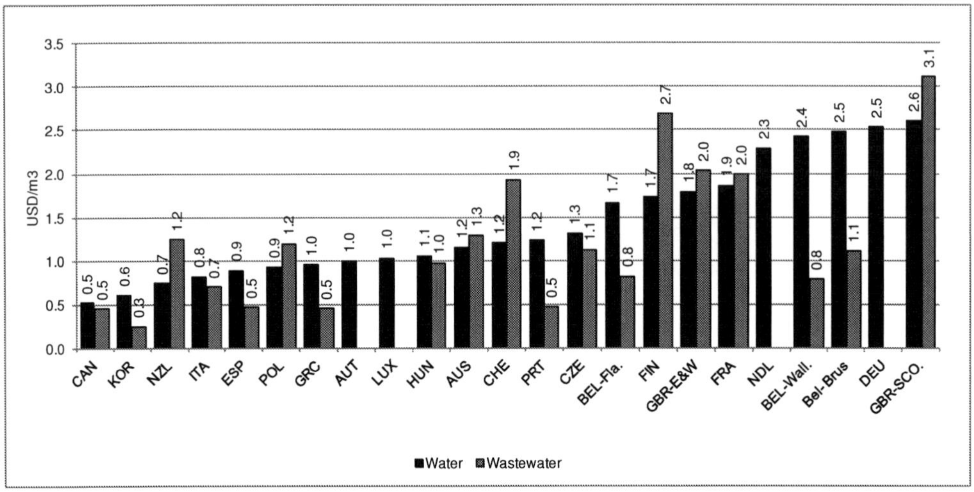

Notes: The sum of the unit values may not match the total unit price of water supply and sanitation indicated in Figure 2.1, as Figure 2.1 data are mostly based on national averages communicated by the countries.

Source: OECD estimates based on public sources validated by the countries.

Table 2.4 reports changes in unit prices of water supply and sanitation services in OECD and selected non-OECD countries over various periods. It shows a significant increase. In most cases in OECD countries, this was driven primarily by increases in wastewater charges, as the latter were brought in line with the costs of investment needed for environmental compliance. In some countries (Australia, Italy, England and Wales, and the Czech Republic), water tariffs increased at nominal rates that were twice the Consumer Price Index (CPI) or larger, sometimes linked with the introduction of private sector participation to spur investment that had been languishing in previous years. In some countries (France, Germany and Scotland), recent increases are slower than in previous periods.

Table 2.4. **Household water and wastewater tariff changes in OECD and selected non-OECD countries, 2008**

	Real average annual rate of change, %[1]			
	Period	Water	Wastewater	Total water and wastewater[2]
Australia	2003-07	2.69	1.85	2.24
Belgium	..	..	..	..
Wallonia	2005-06	1.98	17.42	5.37
Canada	1999-04	-0.32	5.58	2.18
Czech Republic	2000-07	3.31	3.54	3.41
Denmark	2000-06	..		1.67
Finland	2000-08	1.26	2.29	1.88
France	2000-05	0.07	4.29	2.12
Germany	2000-07	-0.63	..	..
Greece	2000-06	-0.96	-0.52	-0.82
Hungary	2000-05	2.65	5.82	4.1
Iceland	..	..	..	..
Ireland	..	..	..	..
Italy	2005-07	2.44	4.41	3.33
Japan	1999-03		..	0.24
Korea	2000-06	1.23	7.39	2.79
Luxembourg	1994-99	0.34	..	..
Mexico	2006-07	..	..	3.43
Netherlands	2000-07	-1.33	..	..
New Zealand[3]	2003-07	-6.11	-6.52	-6.37
Portugal	2004-07	0.14	-0.36	0.00
Spain	2000-06	0.74	10.24	3.37
Sweden	2000-08		..	0.48
United Kingdom:	..	..	..	..
England and Wales	2001-06	2.73	2.98	2.87
Scotland	2004-07	0.41	0.39	0.41

Table 2.4. **Household water and wastewater tariff changes in OECD and selected non-OECD countries, 2008** *(continued)*

Nominal average annual changes in unit revenue, (%)		
	Period	Water and wastewater
EECCA		
Azerbaijan	2000-02	12.35
Kazakhstan	2000-06	10.33
Kyrgyz Republic	2000-05	-7.79
Moldova	2000-06	7.19
Russian Federation	2003-05	26.80
Tajikistan	2000-05	2.86
Ukraine	2000-07	14.46
Uzbekistan	2003-07	0.34
BRIICS		
Brazil	2002-06	18.67
China	2003-07	11.68
India		..
Indonesia	2001-04	22.05
Russian Federation	2003-05	26.80
South Africa	2002-06	17.05

Notes:

1. See Table A.3 (in Annex A) for details on the underlying tariff data.

2. Data sources for water, wastewater and total are from different sources for Wallonia and Canada.

3. Data refer only to Metrowater Utility (Auckland).

Source: OECD estimates based on: (i) for OECD countries: country replies to the OECD 2007-08 Survey (when time series were provided) or other sources validated by the countries. Underlying tariffs in USD were computed using the official OECD exchange rate of the relevant years; (ii) for non-OECD countries: IB-Net data (2007 or most recent year).

The increases in water supply and sanitation tariffs were sometimes driven by an increase in value-added tax (VAT) and other taxes and fees (see Annex A for a synthetic presentation). In addition, in many countries domestic water bills include abstraction charges, as well as a variety of other taxes. Other taxes can be comparable or even higher than VAT (*e.g.* in the Netherlands, the tap water tax amounts to over 10% of the average domestic water charge). Similarly, pollution charges are explicitly included in the wastewater bill in the Flanders and Brussels regions in Belgium, in Canada, Denmark, France, Germany, Hungary, Portugal and Spain. Other taxes are levied in Denmark, France and Spain.

The considerable differences across countries further complicate cross-country comparisons of tariff levels. Different economic signals are sent to both consumers and producers: VAT and other taxes can affect final demand and the affordability of services, but do not contribute to cost recovery.

Tariff structures for water supply in OECD countries

This section discusses trends in the adoption of different tariff structures, their diversity across and within countries, and the composition of WSS bills.

Information collected from the surveys is synthesised in Table 2.5. This information is not comprehensive, as tariff structures are often decided at local the level and are not consistent within a country; other structures could be found in countries.

The situation presented in Table 2.5 is very similar to that already highlighted in the previous OECD survey. The main difference is a smaller number of countries where the use of flat fees and decreasing block tariff structures were detected. This relative inertia can be explained by the complexity attached to changing tariff structure. Moving from flat fees to volumetric rates, for instance, entails the installation of meters; and changing the structures of volumetric tariffs (*e.g.* changing the number or size of blocks in an increasing-block tariff [IBT] scheme) requires that impacts are assessed on the financial sustainability of the provider, the affordability for different consumer groups and short- and long-term impacts of demands. It also requires extensive consultation with the public.

The data confirms that in numerous OECD countries, multiple structures coexist: basically all types of tariff structures are used in different areas of Canada and the United States. Even in countries where IBTs prevail, their design varies across providers in terms of the number and size of blocks. The diversity of tariff structures in a country is generally linked with the degree of decentralisation of the tariff-setting process. In Austria, each municipality determines its own structure, based on the cost recovery principle and taking into consideration social factors as well as geographic and climatic conditions. In Mexico, tariff structures are set by each municipality, according to the laws that apply to each federal state.

Decentralised decisions concerning tariff structures can coexist with centralised controls over tariff levels. This is the case in the Czech Republic, where tariffs are set by individual utilities, but are subject to price controls by the Ministry of Finance. Finally, in some countries the regulatory setting at national level defines the basic principles for the determination of tariff structures and levels, including the number of blocks, while leaving some degree of flexibility at local level so that local, technical and social conditions can be taken into consideration. This is, for instance, the case in Portugal (see Box 2.1). Italy is another example where general rules for pricing WSS are set but diversity in their application is significant. The fundamental difference with the Portuguese case is that tariff articulation is not regulated and supervised by a national regulator, but by local regulators at “Ambit” level.

Table 2.5. **Household tariff structure for drinking water supply in OECD countries, 2008**

	Connection fees	Flat fee	Types of tariff structures[1] Constant volumetric B1	B2	B3	Increasing block tariffs C1	C2	C3	Fixed element base
Australia	..			x			x		..
Austria	..			o					..
Belgium									
Flanders	..						x		..
Brussels	Yes						x		..
Wallonia	Yes						x		..
Canada	Yes	x	x	x	x	x	x	x	Mainly property value
Czech Republic	Yes[2]	x	o						Pipe/meter size, user group
Denmark	Yes		x	x					Area (urban/rural)
Finland	Yes			x					Meter size
France	Yes			x					..
Germany[3]	..			x					Meter size
Greece	..			o			o		..
Hungary	Yes		o						..
Italy	Yes						x		..
Japan	..						o		..
Korea	Yes						x		Meter size
Luxembourg	..			o					..
Mexico	..						o	o	..
Netherlands	Yes			o					..
New Zealand	..			o					..
Norway[4]	Yes			o					..
Poland	..		o						..
Portugal	Yes						x		..
Slovak Republic	..		o						..
Spain	Yes						x	x	Meter size
Sweden	Yes	x	x	x	x				Meter size
Switzerland	Yes			x					..
Turkey	..		o			o			..
United Kingdom	..								..
N. Ireland	Yes	x							Property value
England and Wales		o		o					..
Scotland	..	x							Property value
United States	..		o	o		o	o		..

1. Tariff types:
   B1 Constant volumetric rate with no fixed charge
   B2 Constant volumetric rate + fixed charge
   B3 Constant volumetric rate + a minimum charge + fixed charge
   C1 Increasing block tariffs with no fixed charge
   C2 Increasing block tariffs + fixed charge
   C3 Increasing block tariffs + a minimum charge + fixed charge
2. GWI 2008 database.
3. Federal Association of the German Gas and Water Industries (BGW) data.
4. Kostra database.

Source: OECD 2007-08 Survey and GWI survey: x=OECD survey data; o=GWI data.

Box 2.1. National principles, national regulation, local solutions: the case of Portugal

In Portugal, 275 municipal monopolies provide retail services, while 15 regional companies provide wholesale services at multi-municipal level. In 2007, the economic regulator (IRAR; ERSAR, as of October 2009) assessed the financial conditions of wholesale providers and found that one-third was in "challenging" conditions, while another third was "unsustainable". The analysis concluded that, while structural factors such as scale and population density play a role, the financial bottleneck occurred at the retail level. The main reasons were that retail customer tariffs remained below cost recovery levels, particularly for wastewater, while municipal funding for investment faced competition from other sectors.

IRAR identified the heterogeneity of retail tariffs as a source of confusion for the customers, which opened the door to lengthy and politicised tariff review processes with little buy-in from the population and, consequently, hindered willingness-to-pay. To overcome these difficulties, IRAR proposed a new legislation reflecting a tariff strategy that sought to ensure an adequate allocation of decision-making prerogatives. To this end, some key principles (*e.g.* a progressive move towards full cost recovery, affordability for those in need, transparency) and some tariff features are now set by law. The regulator also issues non-binding recommendations and takes on a coaching role to help municipalities define locally adapted solutions. Significant discretion is left to local authorities who approve water service tariffs. In particular, each municipality can determine the tariff levels for each block, and is required to define a social price plan for poor households.

Other aspects that are clarified at national level include end user and operator rights and duties; minimum information requirements and terminology for invoices, meter reading periodicity, invoicing frequency and payment terms; possibility of seasonal variable water tariffs (peak-load pricing); activities that are included in the tariffs *vs.* auxiliary services that can be charged specifically; and cost items to be included in the determination of service costs.

Source: Pires, J.S. (2007), "Consumer Tariffs in Practice: The Portuguese Experience", presentation made at the OECD Expert Meeting on Water Pricing and Financing, 14-15 November, available at *www.oecd.org/water*.

The basis for computing recurrent fixed charges also differs across countries, although in most cases meter and pipe size, property size or property value are cited as guiding factors. The determination of fixed charges can also vary across different providers within a country. An emerging trend in some OECD countries is the increasing use of fixed charges alongside volumetric components, or the progressive increase in the weight of fixed charges in the overall bill. This trend is driven by providers' attempts to improve financial sustainability by: (i) reflecting in the bill the structure of the cost of water provision, which includes a large share of fixed recurrent costs (*e.g.* billing and collection); and (ii) reducing the impact on revenues of decreasing sales due to the progressive decline in water use intensity.

Tariff structures for sanitation in OECD countries

Increasingly, separate wastewater charges are being introduced to recover wastewater management costs. As for water supply, there are differences both across countries and across providers within countries in the determination of tariff structures, including connection and fixed charges.

Most of the countries that responded to the OECD 2007-08 Survey used the same tariff structures for wastewater as for water supply services, often combining a fixed and a variable element (see Table 2.6). Their levels, however, generally differ from those of drinking water tariffs, and so can the number of blocks used when increasing or decreasing block tariffs are used. In most cases the variable wastewater charge is applied to the volume of water used (or a percentage thereof, as in the case of Northern Ireland), or a percentage of the variable water charge.

Most countries levy separate charges for sewerage *vs.* wastewater treatment, although in most cases the basis for charging remains water consumption; only the size of the volumetric rate differs. In some cases (*e.g.* Belgium, Denmark, Italy and Sweden), customers receive a combined bill for drinking water, sewerage and sewage treatment services. Separate invoices, or separate information on one single bill, are adopted in Australia, Canada, Finland, France, Germany, Hungary, Korea, the Netherlands, the United Kingdom and the United States.

Pricing water supply and sanitation for industrial users

This section reports the main data collected on prices for water supply and sanitation services provided to industrial users in OECD countries. It essentially covers tariff structures.

Data collection is more difficult for industrial water and wastewater tariffs. Differences across productive sectors, for instance, provide an additional layer of complexity. For the OECD 2007-08 Survey, it was decided not to collect data on average tariff levels, but rather focus on: (i) industrial tariff structures and the basis for their calculation; and (ii) cost recovery levels from industrial users.

Table 2.7 summarises the information that was received concerning water tariffs, while Table 2.8 focuses on wastewater collection (S) and wastewater treatment (ST) charges. The additional information collected through the 2008 GWI survey is presented in the text.

Table 2.6. **Domestic wastewater charge structures in selected OECD countries, 2008**

	Connection fees		Same tariff structure as water	Separate charge for sewerage (S) and sewage treatment (ST)	Determination of sewerage (S) and sewage treatment (ST) charges: Description
	In place	Same as water connection			
Belgium					
Flanders	Yes	No	No	Yes	Uniform rate (different for S and ST) applied to water consumption
Brussels	Yes	..	No	Yes	Water use
Wallonia[1]	..	..	Yes (C2)	No	Fixed and volume-based (based on true-cost of water supply and sanitation).
Canada	Yes	..	..	..	..
Czech Republic	No	-	Yes (Flat fee)	No	Water use, plus revenues and taxes
Denmark	Yes	Yes	No	Yes	Water use
Finland[2]	Yes	No	Yes (B2)	No	Water use
France	..	..	Yes (B2)	Yes	..
Hungary	Yes	No	Yes (B1)	..	..
Italy	Yes	..	Yes (C2)	No	Water use
Korea	..	..	No	Yes	Water use
Mexico	Yes	Yes	..	..	..
Portugal	..	..	No	Yes	Based on water use, defined as % of water tariff
Spain	..	..	Yes (C2, C3)	Yes	Based on water use, defined as % of water tariff
Sweden	Yes	No	Yes (Flat fee, B1, B2, B3)	Yes	The same quantity as for water
Switzerland	Yes	No	No	Yes	S: Connection fee + fixed charge ST: Volumetric charge
United Kingdom	..	..	..	..	..
N. Ireland	Yes	No	Yes (Flat fee)	No	Total costs for wastewater service apportioned according to wastewater produced which is 95% of water consumed
Scotland	Yes	No	Yes (Flat fee)	Yes	Unmetered charges based on Council Tax bands

1. CVD (*Coût Vérité Distribution*): water distribution cost determined by each utility in accordance with the government budgetary plan of the water sector. CVA (*Coût Vérité Assainissement*): water treatment cost fixed for the region by the *Société Publique de Gestion de l'Eau* within the framework of its management contract
2. Communication from the Ministry of the Environment.

Source: OECD 2007-08 Survey.

Table 2.7. **Tariff structure for industrial water services from the public system in OECD countries, 2008**

	Connection fees	Connection fees	Fixed element base: Description	Types of tariff structures[1]	Flat fee	Constant volumetric			Increasing block tariffs			Decreasing block tariffs			Full cost recovery	
	1999	2008		1999	A	B1	B2	B3	C1	C2	C3	D1	D2	D3	1999	2008
Australia			..	B2			x			x					Yes	
Austria			..	B2											Yes	
Belgium			..	B2											Yes	
Flanders		Yes	..	..									x			Yes
Brussels		Yes	..	..									x			Yes
Wallonia		Yes	..	..						x			x			Yes
Canada		Yes	..	A, B, D	x	x	x	x	x	x	x	x	x	x	No	No
Czech Republic		No	..	..	x										Yes	Yes
Denmark	Yes	Yes	..	B2		x	x								Yes	Yes
Finland	Yes	Yes	Meter size	B2			x								Yes	Yes
France	Yes	Yes	..	D2										x	Yes	No
Germany		Yes	Meter size	B2			x								Yes	Yes[2]
Greece	Yes	Yes		B1											..	
Hungary				B1											..	Yes[3]
Ireland	Yes			B1											No	
Italy		Yes		B2						x					No	No
Japan				B2											No	
Korea		No	Meter size	B2						x					Yes	No
Luxembourg			..													
Mexico			..	C2						x					No	Yes
Netherlands	Yes		Meter size	B2											Yes	
New Zealand			..	B2											..	..
Norway	Yes	Yes	..	B1											Yes	
Portugal		Yes	Meter size	C2						x					No	..

Table 2.7. **Tariff structure for industrial water services from the public system in OECD countries, 2008** *(continued)*

	Connection fees	Connection fees	Fixed element base:		Types of tariff structures[1]											
					Flat fee	Constant volumetric			Increasing block tariffs			Decreasing block tariffs			Full cost recovery	
	1999	2008	Description	1999	A	B1	B2	B3	C1	C2	C3	D1	D2	D3	1999	2008
Spain		Yes (higher)	Meter size	C2, other						x					No	No
Sweden		Yes	Property/ meter size	B2		x	x								Yes	Yes
Switzerland			..	B2											..	..
Turkey			..	B2											No	
United Kingdom	Yes		Pipe size	B2											Yes	
N. Ireland[4]		Yes	..				x						x			Yes
United States	Yes			C											Yes	

Notes:

1. Tariff types:
   - A Flat fee
   - B1 Constant volumetric rate with NO fixed charge
   - B2 Constant volumetric rate + fixed charge
   - B3 Constant volumetric rate + a minimum charge + fixed charge
   - C1 Increasing block tariffs with NO fixed charge
   - C2 Increasing block tariffs + fixed charge
   - C3 Increasing block tariffs + a minimum charge + fixed charge
   - D1 Decreasing block tariffs with NO fixed charge
   - D2 Decreasing block tariffs + fixed charge
   - D3 Decreasing block tariffs + a minimum charge + fixed charge
2. Federal Association of the German Gas and Water Industries (BGW), 2008.
3. Environmental Performance Review.
4. Fixed element: Un-metered industrial users based on business value and estimated water used/wastewater produced as per households. Decreasing Block Tariffs for larger users.

Source: OECD 2007-08 Survey.

Table 2.8. **Tariff structure for industrial wastewater collection and treatment in OECD countries, 2008**

	Separate charge for sewerage (S) and sewage treatment (ST)	Same tariff structure as water	Determination of sewerage (S) and sewage treatment (ST) charges: Description	Special tariffs
Australia	..	Yes	..	..
Belgium	..	..	..	..
Flanders	Yes	No	S: Mostly consumption or pollution unit ST: Fixed rate per pollution unit	Yes
Brussels[1]	No	No	S: Based on consumption ST: Based on pollution	Yes
Wallonia[2]	Yes	Yes, for the domestic part	Based on pollution: EUR 9 for unit of collection and treatment	No
Canada[3]	..	..	..	Yes
Czech Republic	No	Yes	Based on water consumption	No
Denmark	Only for large users	No	S and ST: Based on water consumption ST large users: Based on pollution/consumption	Yes
Finland	No	Yes	Mostly based on water use	Yes
France	..	No	..	No
Italy	Yes	No	Mostly on pollution content and type of industrial activity	Yes
Korea	Yes	No	Based on water use	Yes
Luxembourg	..	..	..	..
Mexico	Yes	Yes	S: Percentage of the water tariff ST: Percentage of the discharge conditions	Yes[4]
Portugal	Yes	No	Mostly water use based	No
Spain	Yes	No	Water used plus fixed charge per type/quantity of pollutant	Yes
Sweden	Yes	Yes	Based on water use	Yes
Switzerland	Yes	No	Based on pollution	Yes
United Kingdom	..	..	..	..
N. Ireland	No	Yes	Based on water consumption	Yes
Scotland	..	..	..	..
United States	..	..	..	..

Notes:

1. Tariffs vary by municipalities in the three regions. Maximum tariff is set by public authorities. Pollution unit = unit of pollution load. Pollution units are determined based on the analysis of discharged wastewater or by means of conversion coefficients.
2. Tariffs can be calculated according to two methods: a flat rate based on the type industrial activity, or a real price based on analyses of waste water.
3. Industries must pay CVA on the part of the water consumption dedicated to domestic use.
4. Tariff structures vary by municipalities.

Source: OECD 2007-08 Survey.

Compared to tariff structures for household, a few more countries and regions use decreasing block tariffs, particularly for large users. The objective of keeping large customers that provide substantial revenues and regular flows seems to inhibit the use of tariff structures that may provide incentives to reduce water use.

With regards to wastewater management, the trend is towards the increasing use of separate charges for wastewater collection vs. wastewater treatment, with the latter increasingly based on the pollution load of industrial effluents. This indicates an increasing willingness to charge industrial users for the actual treatment costs. Sewerage charges, on the other hand, continue to be mainly volumetric rates (sometimes differentiated by sector) applied to all or a share of the volume of water consumed, and sometimes accompanied by a fixed charge. In many countries, special tariffs are also in place.

Connection charges for households and industrial users

The computation of one-time connection charges for water supply and sanitation services is the area where the most marked differences were found across providers in terms of criteria used and levels applied. In many cases there are no uniform charges even for a single provider, and these are determined on a case-by-case basis, based on the actual cost of connecting the customer. This *ad hoc* calculation is sometimes reserved for industrial users and sometimes only to the larger ones (*e.g.* some cities in the United States, Korea).

Connection charges can vary across broad ranges for the same provider. For example, in Indianapolis, water connections cost between USD 19 and USD 337, while wastewater connections are USD 2 500. On the other hand, in New York City both water and wastewater connections cost USD 200, with an additional "water tap fee" of USD 258. Fees of the same order of magnitude or higher (or much higher, as in Korea) apply in Australia, Belgium, the Czech Republic, Hungary, the Netherlands and the United Kingdom. In Canada, Portugal and Spain, the levels are one order of magnitude lower in the GWI sampled cities.

A synthesis

In this section, some of the key trends identified over the last decade for OECD countries are highlighted, based on a comparison of 1999 and 2008 data.

OECD countries increasingly use abstraction and pollution charges. In some countries, pollution charges are increasingly tailored to environmental challenges and used as an incentive to reduce pollution. However, in most cases, these instruments tend to be used as sources of revenues. They are not designed to cover the scarcity value of the resource; this can primarily be explained by the administrative costs of building and administering the information base and computing capacity to calculate this value.

As regards water supply services to households:

- Real prices have tended to increase, at times substantially, both in OECD and non-OECD countries. This may signal an increased role of tariffs in cost recovery.
- Charges for water supply are increasingly severed from charges for sanitation services. The latter tend to be computed based on actual costs of compliance with effluent standards. This sometimes leads to substantial increases in the price of wastewater services.
- The structure of water supply and sanitation tariffs tends to change: fewer countries report using decreasing block tariffs and flat-fee systems; these systems are changed for two-part tariffs combining fixed charges and variable fees; variable fees are either uniform or based on increasing blocks.

As regards water services for industrial uses, decreasing block tariffs for water supply are only used in a few OECD countries. Special charges for wastewater increasingly incorporate the level of pollution, thus reflecting the costs of treatment or of the externality. At the same time, there is a tendency for large industrial consumers to go off-network and to carry out abstractions directly.

Taxes are applied to water bills more often. They vary greatly across countries and make cross-country comparisons difficult. This situation results in contrasting signals being sent to users. Typically, VAT and other taxes can affect final demand and the affordability of service, but do not contribute to cost recovery.

Notes

1. This was later merged with a complementary database including information on other economic instruments and voluntary approaches. This database is now jointly managed by the OECD and the European Environment Agency (EEA). See *www.oecd.org/env/policies/database*.

2. Not all countries levy abstraction charges on municipal water providers, and retail tariffs do not always contain such charges. Therefore, final users of WSS services may not receive any signal concerning water resources, their scarcity, or the cost of water resource management (WRM) activities. One country where final users receive this signal is Korea, where laws were passed in the period 1999-2002 to set up river basin authorities financed by bulk water use charges, which are passed on to consumers through tariffs.

3. This information was provided by national experts as part of the response to the OECD 2007-08 Survey.

References

Please also see Annex B for additional sources of information per country.

BGW (German Water Association) (2008), "Profile of the German Water Industry", *www.dvgw.de/fileadmin/dvgw/wasser/organisation/branchenbild2008_en.pdf.*

Ecologic (2008), "Water Pricing and Costs of Services in Germany".

Gaulke, A. (2007), "Water Abstraction Fees in Germany", presentation made at the EEB Seminar, Brussels, 6 October.

NSW Independent Pricing and Regulatory Tribunal (2004), "Bulk Water Prices from 2005/06", Issues Paper.

OECD (1999), *The Price of Water: Trends in OECD Countries*, OECD Publishing, DOI: *http://dx.doi.org/10.1787/9789264173996-en.*

OECD (2003a), *Social Issues in the Provision and Pricing of Water Services*, OECD Publishing, DOI: *http://dx.doi.org/10.1787/9789264099890-en.*

OECD (2003b), *OECD Environmental Performance Reviews: Austria 2003*, OECD Publishing, DOI: *http://dx.doi.org/10.1787/9789264018891-en.*

OECD (2006), *OECD Environmental Performance Reviews: Korea 2006*, OECD Publishing, DOI: *http://dx.doi.org/10.1787/9789264024045-en.*

OECD (2008a), *OECD Environmental Performance Reviews: Hungary 2008*, OECD Publishing, DOI: *http://dx.doi.org/10.1787/9789264049284-en.*

OECD (2008b), *OECD Environmental Performance Reviews: Australia 2007*, OECD Publishing, DOI: *http://dx.doi.org/10.1787/9789264039612-en.*

OECD (2008c), *OECD Environmental Performance Reviews: Denmark 2007*, OECD Publishing, DOI: 10.1787/9789264039582-en, *http://dx.doi.org/10.1787/9789264039582-en.*

OECD (2010), *Sustainable Management of Water Resources in Agriculture*, OECD Publishing, *www.oecd.org/tad/env.*

Pires, J.S. (2007), "Consumer Tariffs in Practice: The Portuguese Experience", presentation made at the OECD Expert Meeting on Water Pricing and Financing, 14-15 November.

Rees, C., J. Winpenny and A. Hall (2008), "Water Financing and Governance", *GWP TEC Background Paper* 12.

Chapter 3

Policy challenges related to water pricing

This chapter focuses on two objectives of pricing for water supply and sanitation services: the financial sustainability of services, and the affordability of water bills. It shares data on cost-coverage ratio for domestic water supply and sanitation services and on the share of disposable income allocated to water bills in OECD countries. Data indicate that affordability can be an issue for segments of the population.

The data presented in the previous chapter is analysed with a view to document how the policy dilemmas identified in Chapter 1 are being addressed by OECD countries. The focus is essentially on cost recovery and on affordability. Before we turn to this, let us first discuss the new context for water pricing, as this reinforces the dilemma sketched out above.

A new context for water pricing

As mentioned in Chapter 1, two main sets of challenges face the water sector in OECD and non-OECD countries. One regards the increasing competition for the use of water resources for human consumption, productive uses, and the support of ecosystems. Limited water resources availability, their deteriorating quality, the impacts of climate change and poor management, all contribute to the problem, which is compounded by the potential destructive impact of water-related extreme events (droughts and floods).

The other challenge is the need to ensure access to adequate, sustainable and affordable water and sanitation services for all. While in some regions water scarcity is one of its limiting factors, this goal is again mainly constrained by management factors including ill-conceived investment, the deterioration of infrastructure due to insufficient cash flows (which in turn limit access to external finance), inappropriate regulatory frameworks, and limited managerial and regulatory capabilities.

These challenges set the scene for pricing water-related services (both water resource management and water supply and sanitation). They confirm that water policies need a mechanism to allocate water where it is most needed, and a financing mechanism to generate revenues and leverage additional sources of finance. It is noteworthy that recent analyses confirm that these challenges are not limited to developing countries: OECD countries are facing similar (although different) issues, and efficient responses require, *inter alia*, that water-related services are priced in ways which contribute to a range of environmental, social, economic and financial objectives.

The water resource management challenge

Water scarcity is an increasing threat in many countries and regions, as water pollution and overuse reduce available sources, and population and economic growth increase competition between different uses. Some regions, such as parts of South and East Asia, Australia, Africa and the Middle East, do suffer from actual physical water shortages. According to the 2006 World Water Development Report (UNESCO, 2006), in 2000 20% of the world's population had no appreciable natural water supply; 65% had low-to-moderate supply with 50% of global run-off and only 15% lived with relative abundance with the remaining 50% of global run-off.

But scarcity is not a mere physical phenomenon. "Dry" areas may not be water scarce if use remains within the limits of local availability. Conversely, "wet" areas may be stressed if use approaches the limits of availability. Currently, 1.4 billion people live in basins where water usage exceeds recharge. For example, the Middle East/North Africa region is running at a deficit, as annual withdrawals exceed annual resource replenishment.

A different definition of water stress, which uses the percentage ratio of withdrawals to available resources, is better able to reflect different geographical, economic and cultural circumstances. The OECD's water stress indicator is based on the ratio of annual water withdrawals to annual water availability. Below 10% water stress is low; the 10-20% range indicates moderate stress and suggests that: "water availability is becoming a constraint on development and that significant investments are needed to provide adequate supplies"; above 20% stress is medium and "both supply and demand will need to be managed and conflicts among competing uses will need to be resolved"; finally, when annual withdrawals exceed 40% of annual water availability, stress is severe (OECD, 2006a).[1]

The most recent OECD data (see OECD, 2008a) shows that Italy and Korea are already under severe stress with freshwater abstractions exceeding 40% of renewable resources, while Belgium and Spain exhibit moderate stress and seven more countries are in the upper echelons of moderate stress, calling for increased investment in resource development.

Looking to the future, the water imbalance situation appears to be worsening. The *OECD Environmental Outlook to 2030* (OECD, 2008b) projects that by 2030 the number of people living in areas affected by severe water stress will increase by another 1 billion from the 2005 baseline to an estimated 3.9 billion people (47% of the world population), mostly in non-OECD countries.

Increased demand for water is commonly attributed to population increases, but other factors (*e.g.* economic development, migration, urbanisation) play an important role. Over the last century, water use has increased six-fold while population has increased at only a half of that rate (Bergkamp and Sadoff, 2008). Changes in patterns of withdrawal from nature and consumption are therefore also a major part of the problem.

Overall abstraction in OECD countries has remained largely stable since 1990. While this may have partly resulted from policies that favoured more efficient water use and its decoupling from economic and population growth, in some countries demand has had to shrink in response to drought. In addition, this data hides significant variations at the geographical and sector level. For instance, gross water abstractions increased quite substantially in Korea and Turkey over the period.

Indicators of water resource use intensity show great variations among and within individual countries. The national indicator may thus conceal unsustainable use in some regions and periods, and high dependence on water from other basins. In arid regions, freshwater resources may at times be limited to the extent that demand for water can be met only by going beyond sustainable use in terms of quantity. At world level, it is estimated that water demand has risen by more than double the rate of population growth in the last century. Agriculture is the largest user of water world-wide, accounting for about 70% of the total global freshwater withdrawals; abstractions for irrigation are estimated to have increased by over 75% since 1960 (OECD, 2008c).

Water use is projected to increase at a much higher pace in developing countries, where agriculture is by far the main user, resulting in a global share of agriculture water use of about 70%.[2] According to OECD (2008b), agricultural production will increase twice as fast in developing countries than in OECD countries, and electricity and industrial production are also projected to grow much faster in non-OECD countries.

The water scarcity threat is compounded by the continued deterioration of freshwater quality from both point and non-point sources. While most OECD countries have tackled surface water pollution problems, mainly by regulating discharges from large point sources and investing in municipal wastewater treatment (reaching a population coverage rate that is about 70%), pollution from diffuse sources continues, particularly from agriculture.

Climate change is also expected to affect the capacity of water systems to meet human and other needs while preserving resource quality and availability. Climate change's main water-related impacts are expected to be in terms of shifting and more variable hydrological regimes, *i.e.* changes in water distribution around the world, changes in its seasonal and annual variability, and an increase in the frequency and/or intensity of extreme events (EEA, 2007). Rising sea levels at one end of the water-flow cycle threaten the world's mega-deltas, while up-stream the vast populations dependent on glacial melt (one-sixth of the world population) are losing their "water towers": the high altitude glacial reservoirs.

The development path chosen by a country, human consumption patterns for both water and food, and the capacity of governments to set up water resources management systems to guide the allocation of water resources across uses (with due consideration for environmental uses), will determine a country's capacity to deal with its water scarcity challenges. This confirms that it is critical to allocate water where it is most needed, and to preserve or restore the quality of the resource.

The water and sanitation challenge

Water supply and sanitation raises specific issues. These services have a strong "general interest" component, as their consumption generates benefits not just for the individual, but for the community as a whole. Recognising this, in 2002 the Committee on Economic, Social and Cultural Rights of the United Nations formally recognised the right to water as a human right, and the international community committed to the Millennium Development Goals (MDGs) that aim to halve the proportion of people without access to safe drinking water and basic sanitation by 2015. OECD countries are committed to working with developing countries to achieve this.

The situation is contrasted between OECD and developing countries. Each situation has financial consequences, which should be taken into account when reflecting on pricing water supply and sanitation services.

OECD countries

In most OECD countries, 100% of the population have access to safe drinking water. With few exceptions, water supplied to the main centres is bacteriologically safe (OECD, 2007a). However, in some countries (*e.g.* Mexico, New Zealand, Poland and the United States), a part of the population is not connected to public water supplies. Moreover, quality monitoring of smaller drinking water supplies needs to be improved (OECD, 2003a, 2006a).

Wastewater collection and treatment coverage in OECD countries is also good compared to worldwide levels. However, there are substantial differences in terms of coverage and type of treatment across member countries. Further investment is required to reduce environmental pressures on water resources.

It follows that significant investments will be required to rehabilitate existing infrastructure, bring it into conformity with more stringent environment and health regulation, and maintain service quality over time. OECD (2006b) estimated that investment in water infrastructure to 2030 will need to be larger than that for the transmission and distribution of electricity. France and the United Kingdom will have to increase their water spending as a share of gross domestic product (GDP) by about 20% just to maintain water services at their current levels, while Japan and Korea may have to increase their water spending by more than 40%.

Developing countries

Progress towards achieving the water supply and sanitation (WSS) targets of the MDGs has been overall disappointing. Overall the targets for access to drinking water may be met, but not in all regions/countries, and particularly not in Sub-Saharan Africa. From 2004 to 2006, the number of people without access to drinking water declined by 20%,[3] but 880 million people still lack access, 84% of whom are in rural areas. The situation is even worse for sanitation, as the world is not on track to meet even the limited objective foreseen by the MDG target. Between 1990 and 2006, the proportion of people without improved sanitation decreased by only 8% (from 2.6 to 2.5 billion). At the current rate, the world will miss the MDG sanitation target by over 700 million people. Southern Asia and Sub-Saharan Africa are especially off track in terms of sanitation coverage.

These trends are important from a financial point of view. Various estimates of investment needs to meet the MDGs in developing countries were produced since the Millennium Declaration. The most recent estimates of financial needs for the sector by World Health Organization (see Hutton and Bartram, 2008; Prüss-Üstün *et al.*, 2008) include both investment needs to increase access, as well as recurrent expenditure and investment needed to continue providing services to those that are currently served. It concludes that in 2005-14, annual expenditure requirements in developing countries amount to USD 72 billion, divided as follows: USD 18 billion to increase coverage and USD 54 billion to maintain and renew existing facilities for served populations.

In the future, emerging and developing countries will also face increasing recurrent costs linked with their expanded systems. With the construction of wastewater management networks for an increasingly urbanised population, the cost of sewerage and wastewater treatment will also escalate (in a mature networked system, these costs normally exceeds that of water supply). It is therefore necessary to consider the full set of financial implications of extending first-time access, particularly to networked services.

Ecological sustainability

Pricing water supply and sanitation services can contribute to ecological efficiency if it is used to manage demand (to encourage a more rational and efficient use of the resource) and to recoup the costs of the damages born by the environment (*i.e.* negative impacts on ecosystems, including pollution).

A variety of pricing levels and tariff structures can be used, which contribute to ecological sustainability. Non-uniform flat rates (in use in the United Kingdom and in the former Soviet Union) can be based on features which relate to water consumption (*e.g.* efficient appliances, water reuse). Uniform volumetric rates contribute to ecological sustainability, especially if there is no fixed charge; it requires that households are metered. Increasing block tariffs are even more favourable, especially if marginal rates in the upper blocks are high.

These features have to be kept in mind when considering the other objectives of water pricing policies.

Economic efficiency: a rationale for second-best solutions

As mentioned in the Chapter 1, the economic literature on water pricing generally recognises long-term marginal cost (MC) pricing as the first-best pricing option. But in reality it is seldom used. IWA (2008) indicates that only Australia reported the use of MC pricing as one of the principles for setting water tariffs. The OECD 2007-08 Survey reported that only Italy and Mexico mention MC pricing as a guide to set tariffs, for industrial users.

This may derive from the fact that MC pricing may conflict with some policy objectives (the financial sustainability of operators) and may be costly to operate (in part due to metering).

The OECD 2007-08 Survey confirms that metering is unevenly spread across OECD countries. Countries were asked to provide the share of connections (not only domestic) with functioning meters; the question was asked specifically for individual apartments and single-family houses, to provide for a distinction between metering at the entrance of blocks of apartments and metering of individual flats; data could only be collected for a limited set of countries.

According to the survey, metering penetration is about 63% in Canada (2004 data), 30% in Mexico,[4] 80% in Switzerland, 95% in Sweden (although this figure covers single-family houses only; coverage for individual apartments is reported to be 10%) and about 100% in Belgium, the Czech Republic, France, and Portugal (usually at the level of buildings);

in Ireland, no domestic users are metered (but some 60 000 industrial users are).

Additional data from the literature is available for Denmark (100% of households are metered; see OECD, 2008d), England and Wales (37% of households were metered; see Ofwat, 2009), Australia (100% in the sampled cities), the United States (100% in the sampled cities), Netherlands (96%; Amsterdam is a case in point, with 47% of the population being metered). IWA (2008) indicates the presence of unmetered customers (and the consequent use of flat rates) in Antwerp (Belgium), Amsterdam and Rotterdam (Netherlands), Bergen and Oslo (Norway), Singapore, and all sampled cities in Romania.

It is noteworthy that metering may not be appropriate under every circumstance. It certainly can save long-run costs in terms of reduced impacts on the resource. But metering generates short-term costs (for manufacturing, installation, maintenance, and replacement) and may also reduce revenues for the water company, causing net cash flow problems. Benefits from metering will therefore depend in part on the price elasticity of demand and on the balance between the efficiency gain and reduced long-run cost *vs.* their higher short-term costs (see Herrington, 2006, 2007 and Dalhuisen *et al.*, 2001, 2003, for a discussion).

Financial sustainability: cost recovery for water supply and sanitation services

The financial sustainability of operators (whatever their status) is a requisite for the sustainable operation of water services. Key issues are the level of revenues and their stability or predictability. Because other financing instruments (taxes and transfers) are volatile and beyond the reach of the water community, cost recovery through tariffs is considered a significant driver of the financial sustainability of water operators.[5]

Different tariff structures respond more or less effectively to the financial needs of utilities. From a utility perspective, the large share of fixed costs in WSS service provision would call for their recovery through fixed charges. This is particularly true in those cases where water use is declining (either due to increases in volumetric rates or to structural demographic and economic factors). But such a practice may hinder economic efficiency.

Responses to the OECD 2007-08 Survey on this issue were very sparse for both domestic and industrial users (Tables 3.1, 3.2 and 3.3 respectively). Moreover, the responses may not be comparable as some respondents noted a risk of misunderstanding and one country provided cost coverage ratios that included irrigation services.

For domestic services, providers in most countries on average cover at least operation and maintenance (O&M) costs, sometimes with large margins as in Denmark, England and Wales, and the Netherlands.

Table 3.1. **Cost coverage ratio for domestic water supply and sanitation services in selected OECD countries, OECD survey results, 2008**

	Ratio[1]	Comments
Belgium		
Brussels	1.05	Operating costs most likely include debt service and depreciation
Wallonia	1.11	Operating costs most likely include debt service and depreciation
France	1.0	Operating costs most likely include debt service and depreciation
Portugal	1.29	Estimates refers to municipal concession covering 17% of population
Spain	0.65-0.96	For both urban and irrigation uses; might be higher for urban services
Sweden	1.0	
United Kingdom		
N. Ireland	1.0	Operator is intended to be self-financing

1. Ratio of total WSS (billed) revenues on total operating expenditure.

Source: OECD 2007-08 Survey.

Table 3.2. **Cost coverage ratio for domestic water supply and sanitation services in selected OECD countries, various sources, 2008**

	Ratio[1]	Comments
Australia	1.45	
Canada	1.09	Estimated as total revenues on total O&M expenditure based on utility data
Czech Republic	2.34	Estimated as (volume of water sold)*(consumer W price + consumer WW price) / total O&M expenditure
Denmark	2.78	Estimates based on the unit cost for water only
Germany	1.0	
Hungary	1.09	IB-Net
Italy	1.19	National report (Blue Book, 2008): revenues/operating costs
Mexico	1.16	IB-Net
Netherlands	2.03	Data refer to water supply services only
New Zealand	1.0	IB-Net
Norway	1.49	
Poland	1.16	IB-Net
Slovak Republic	1.42	IB-Net
United Kingdom		
England and Wales	2.62	Ofwat, Financial Performance and Expenditure of the Water Companies in England and Wales, 2007-08 report

1. For IB-Net data, ratio of total WSS (billed) revenues on total operating expenditure.

Source: Based on miscellaneous sources, including IB-Net. If not specified, national data extracted from publicly available sources. The variety of sources makes comparison difficult, if not misleading.

Table 3.3. **Cost recovery levels for industrial water supply and sanitation services from the public system in selected OECD countries, 2008**

	Full cost recovery	Marginal cost
Belgium:	..	..
Flanders	No	No
Brussels	No	No
Wallonia	No	..
Canada	No	No
Czech Republic	Yes	No
Denmark	Yes	No
Finland	Yes	No
France	No	..
Germany	Yes	
Hungary	Yes	
Italy	No	Yes
Korea	No	No
Mexico	No	Yes/with adjustments
Sweden	Yes	No
Switzerland	No	No
United Kingdom	..	..
N. Ireland	Yes	No

Source: OECD 2007-08 Survey.

Based on these data, the operation and maintenance costs of domestic WSS services are generally covered. However, there doesn't appear to be a large margin for operators to also face the need to renew and replace ageing infrastructure, although very few countries provided data on this item. The capacity to cover new investment to extend the system appears questionable; generating revenues to cover full economic or sustainability costs seems to be a remote target only. The picture is similar for industrial users, as full cost recovery through tariffs is far from being the norm and subsidies are still largely used.

An analysis of specific cases suggests that efforts have been made to increase cost recovery from tariffs in many OECD countries. The focus has been primarily on ensuring that effective funding mechanisms are in place to ensure the financial sustainability of the sector, and particularly of wastewater management, where larger investments are needed.

In Finland, the 2001 Water Supply and Sewerage Act requires full cost recovery (including investments) through tariffs in the long run. In Luxembourg, the 2008 Water Law requires that the pricing system be designed to ensure that tariff revenues fully cover at least supply costs, including asset depreciation.

In Switzerland (OECD, 2007b), recovery of operating costs for wastewater services has increased since the introduction of the polluter-pays principle in the 1997 Law on Water Protection. Recent efforts have aimed at improving capital cost coverage based on depreciation, calculated on future replacement value. A special fund for "maintenance of value" was created, with annual contributions calculated on the basis of an average operating life of 33 years for treatment plants and 80 years for sewerage networks. In the canton of Bern, since 2000, an effluent charge is paid by entities discharging in water bodies, including treatment plants operators; revenues go to a "sewerage fund" that is mainly used to finance network extensions, new treatment plants or their extensions.

In Belgium (OECD, 2007c), all regions achieve full cost recovery for drinking water services. In Wallonia, for instance, water supply prices are based on the true cost of water supply (the CVD, for *coût vérité distribution*), which includes the cost of protecting water intake areas. All regions are also setting up arrangements for setting and collecting wastewater charges, which were reformed to achieve a higher degree of cost recovery in the future. Wallonia now has in place a coherent funding mechanism designed to achieve full cost recovery, and wastewater charges are expected to triple over the next ten years. In 2001, Brussels-Capital established a water policy fund to finance the construction of the collector system that takes untreated sewage from the region to the new large treatment facility just north of the capital. The fund is financed mainly by revenue from regional water pollution levies and a contribution from the Flemish Region (which benefits as well).

Affordability of water bills: reconciling different policy objectives

This section analyses affordability measured on income levels. Smets (2008) proposes an affordability index linking water and wastewater bills to disposable income, and provides indications on how to handle data (*e.g.* how to translate *per capita* into household water consumption, depending on the number of household members). The index can be computed at national or at local level, and can vary depending on the type of definition that is given for "poor" households.

Water supply and sanitation bills are computed, using the data collected by the OECD 2007-08 Survey. Incomes are estimated based on household surveys used for the OECD *Growing Unequal?* (OECD, 2008e). Affordability is assessed at the national level (using average net disposable income; see Figure 3.1) and for the lowest decile of the population (using average income of the lowest decile of the population; see Figure 3.2). As for other indicators used in this report, cross-country comparisons of average national affordability levels are fraught with problems and should be interpreted with caution; local affordability estimates can be useful in determining solutions for population groups confronted with an affordability problem even when national averages remain below international thresholds (see the case of Portugal, in Box 3.1).

Figure 3.1. **Water supply and sanitation bills as a share of disposable income in OECD countries, 2008**

Average net disposable income

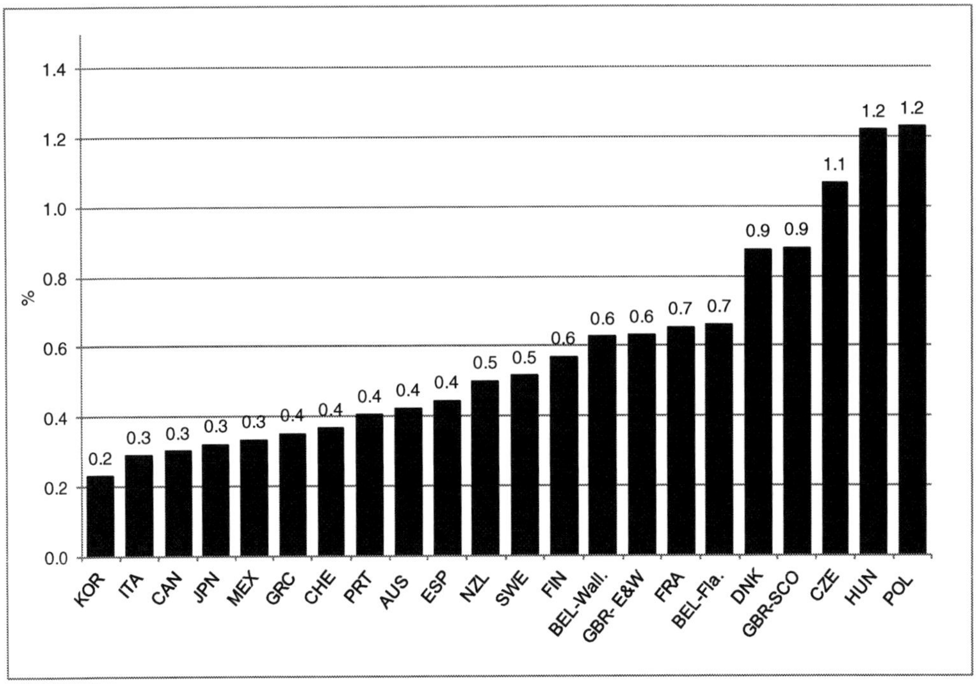

Notes:

Data for water tariffs are calculated for a consumption of 15 m^3/month and based on 2007 PPPs for private consumption.

Data for net disposable income are expressed in USD, 2007 exchange rates and PPPs.

Source: For water bills: OECD estimates based on country replies to the OECD 2007-08 Survey or public sources validated by the countries; for NDI: OECD National Accounts.

Figure 3.2. **Water supply and sanitation bills as a share of disposable income of the lowest decile of the OECD population, 2008**

Average income of the lowest decile of the population

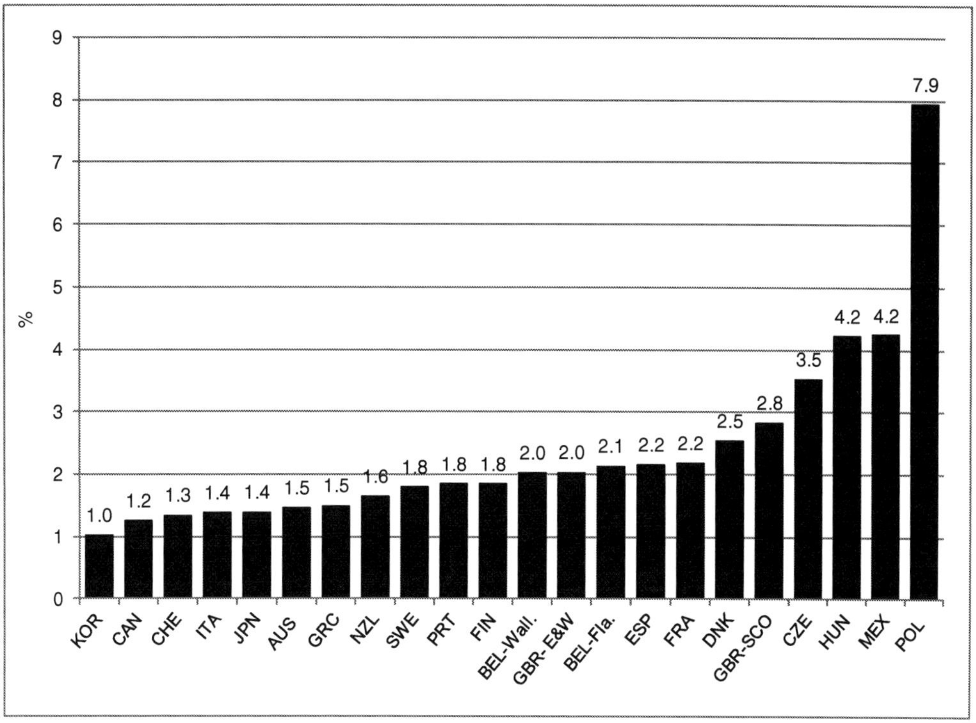

Notes:

Data for water tariffs are calculated for a consumption of 15 m^3/month and expressed in USD adjusted at 2007 PPPs for private consumption.

Data for average income for the lowest decile of the population refer to 2005 and are expressed in USD adjusted for PPPs for private consumption.

Source: For water bills: OECD estimates based on country replies to the OECD 2007-08 Survey or public sources validated by the countries; for NDI: OECD income distribution questionnaire and other OECD databases.

Box 3.1. **Local affordability assessment for tariff policy reform: the case of Portugal**

As part of the process leading to the design of its proposed tariff reform, the economic regulator of WSS services in Portugal (IRAR; ERSAR, as of October 2009) carried out an affordability study. This identified geographically concentrated clusters of population that would fall below the affordability threshold, which had been set at 3% of household disposable income. Some 10.5% of Portuguese households faced bills in excess of the affordability criteria. These were concentrated in 60 out of 309 municipalities in the North and Tagus Valley regions, where 15-30% of households would face unaffordable bills. The affordability study, however also showed that WSS services do not pose an affordability problem for society as a whole, as they represent a very small portion of overall expenditure by household on utility services (including electricity, gas, etc.).

The design of the tariff reform considered these results, by: (i) allowing flexible solutions in different municipalities to address geographically localised affordability problems; (ii) including support from IRAR to local service providers on ways to manage the transition to financial sustainability; and (iii) structuring a communication plan to the public to clarify the real situation with regards to the weight of WSS costs for Portuguese households.

The draft of the new Water Act makes it mandatory for each service provider to draw up a "social plan" which increases affordability by amending the water supply tariff structure for low-income households. As the variable portion of the wastewater bill is set as a certain percentage of the variable portion of the water supply bill, this reduction is also extended to wastewater services. This is to be achieved by mandatory: (i) removal of the requirement to pay the fixed charge; and (ii) extension of the "width" of the first, lowest-priced block beyond that in the standard tariff. Low-income households are defined as those whose annual income is below the national annual minimum wage. Evidence of eligibility will be required, to be renewed every three years – tax returns and statements of government benefits received will be accepted. The precise details of the social plan tariff – in particular the width of the first block – will be determined by the municipality (regardless of whether or not it is the service provider).

Source: Pires, J.S. (2007), "Consumer Tariffs in Practice: The Portuguese Experience", presentation made at the OECD Expert Meeting on Water Pricing and Financing, 14-15 November, available at *www.oecd.org/water*; Herrington, P. (2008), "Portuguese Proposals for Future Household Water Tariffs", report prepared for the 2007-08 OECD Horizontal Water Programme.

Figures show that in OECD countries water supply and sanitation bills do not represent a considerable burden on disposable household income when using average income figures. The share of disposable income households spend on water supply and sanitation bills ranges from 0.2% (in Korea) to above 1% (in Central European countries: the Czech Republic, Hungary and Poland).

The picture is more contrasted when one considers the lowest decile of the population (Figure 3.2). For these groups, average representative bills would represent a significant share of disposable income in Mexico and in some Eastern European countries (Hungary, Poland, and to a lesser extent, the Czech Republic). Problems also appear in countries such as Denmark and Scotland. It is important to note that these figures do not represent what poor households actually pay, as most countries have introduced social tariffs.

A recent survey by Smets (2008) showed that over 50 countries have put in place measures to address affordability problems, while taking into consideration financial sustainability of services. Social tariffs have two separate aspirations: (i) to help the poor deal with affordability constraints; and (ii) to ensure access to a minimum amount of drinking water at very low prices or free of charge.

Two aspects are especially important in determining the success of tariff structures in addressing both financial sustainability and affordability considerations. First, for financial sustainability to be respected, a non-marginal number of users will have to pay more than the long-term average cost of service provision. A second key aspect is the capacity of the structure to appropriately target the poor, defined by Komives *et al.* (2005) as the capacity not to exclude from receiving a subsidy any user that would qualify for it and not providing subsidies to any user that would not quality.

Increasing block tariffs, with a first block provided at zero or very low price, have been set up in a number of countries to guarantee access to a minimum amount of water. This tariff structure does not qualify as a social tariff as it has not been established that poor households consume much less water than more well-off ones: empirical research tends to indicate that income elasticity of water demand is low; in other terms, the poor do not necessarily consume less water than the rich.

While perfect targeting may be too costly to achieve, a number of solutions have been introduced in different countries. One is to adjust tariff structures to account for household size, so as to avoid penalising large families. This solution is being adopted in an increasing number of OECD countries, including the Brussels and Flanders region in Belgium, Luxembourg, and some municipalities in Greece, Portugal and Spain.

A second alternative is a uniform volumetric rate, high enough to meet both financial and economic efficiency objectives; part of the revenues is used to support those families that would otherwise be unable to face their bills. An example is the system adopted in the Wallonia region of Belgium, where water prices have included a charge destined for the Social Water Fund, to be distributed to low-income households.

This confirms that, adequately designed, prices for water supply and sanitation services can be adjusted, both in terms of levels and structures, to contribute to different policy objectives. However, this requires a careful assessment of the possible consequences of prices on ecological, social, economic and financial dimensions; local conditions prevail and should be reflected in the data used for such assessment.

Way forward

Three ways can be envisaged to move the agenda forward with regard to pricing water-related services, building on the OECD 2007-08 Survey and on countries' experience. The ultimate objective would be to help policy makers to improve water-pricing policies, building on best international practices, on lessons from experience, and on robust empirical analyses.

Update the data regularly

The analysis of the recent survey signals that trends in water pricing have changed over the last decade. Moreover, a renewed context is emerging, characterised by daunting environmental, social, economic and financial challenges. The recent macro-economic situation adds more uncertainty: in the short term, some policy packages include investment in the water sector; in the longer term, public moneys will be scarcer, competition to access them will be fierce, and pricing may become an even more topical issue. These factors suggest that there are some benefits to survey pricing practices on a regular basis to identify best practices (*e.g.* on financing renewal of infrastructures, or coping with affordability issues) and learn lessons (*e.g.* on price elasticity of demand, on the effects of specific taxes).

Address data and knowledge gaps

The difficulties encountered in data collection and discussion with a number of country experts provided valuable indications on: (i) the limitations in data availability on key aspects of WSS at the national level; and (ii) difficulties in mobilising and systematising such information from the relevant local entities. Analysis confirmed that cross-country comparisons based on national data is difficult.

This highlights possible avenues for future work aimed at improving data collection at local level, reducing inconsistencies in sampling and aggregation at national level, and creating a set of indicators that may facilitate comparability on policy-relevant variables across OECD countries.

In particular, experience calls for a rethinking of the collection, analysis and presentation of tariff data at local level. A dialogue with country experts could be undertaken with the objective of suggesting some harmonisation in the aggregation and sampling methods, *e.g.* in terms of share of population covered, possible differentiation across geographical areas based on pre-identified characteristics, etc.

Another direction could also be explored to monitor cost recovery performance. The method currently used (including in the OECD 2007-08 Survey) asks for a lot of data and relies on uncertain definitions. It might be easier to collect information about subsidies and transfers (from outside the water sector) received by utilities.

Develop a checklist for policy makers

The discussions above have pointed to a number of practical steps which policy makers may wish to consider while assessing or revising tariffs for water supply and sanitation services, or for water resource management. These considerations could be organised in a checklist for policy makers, which would build on the analysis herein and on best practices in OECD and non-OECD countries.

The checklist could take the form of a sequence of questions which policy makers may wish to consider when designing or revising tariffs for water-related services.

Notes

1. Water stress indicators are imperfect when aggregated at the country level. For instance, in Australia, one-third of the country is arid, one-third semi-arid and the high rainfall areas in the north are far from the density populated areas in the south, which is also where demand from agriculture is concentrated. Any meaningful analysis of stress/scarcity needs therefore to be based on the computation of regional water balances.

2. Globally, 20% of freshwater use is accounted for by industry, and the remainder (5-10%) by domestic use OECD (2006b).

3. Down from 1.1 billion reported for 2004 in WHO/UNICEF (2006).

4. A different source states that 52% of households who have access to the service are metered (INEGI, 2003).

5. The OECD has commissioned research on indicators for financial sustainability. A report is published separately (see Massarutto *et al.*, 2010), which proposes a methodology for assessing the financial sustainability of operators of water supply and sanitation services and applies it in three countries.

References

Please also see Annex B for additional sources of information per country.

Bergkamp, G. and C.W. Sadoff (2008), "Water in a Sustainable Economy", in *State of the World 2008: Innovations for a Sustainable Economy*, Worldwatch Institute, Washington, DC, pp. 107-122.

Dalhuisen, J., *et al.* (2001), "Price and Income Elasticities of Residential Water Demand: A Meta-Analysis", Tinbergen Institute discussion paper, TI 2001-057/3, Vrije Universiteit Amsterdam.

Dalhuisen, J., *et al.* (2003), "Price and Income Elasticities of Residential Water Demand: A Meta-Analysis", *Land Economics*, Vol. 79, May.

EEA (European Environment Agency) (2007), "Climate Change and Adaptation Issues", Technical Report No. 2/2007, EEA, Copenhagen.

Herrington, P. (2006), *Critical Review of Relevant Research Concerning the Effects of Charging and Collection Methods on Water Demand, Different Customer Groups and Debt*, UKWIR, London.

Herrington, P. (2007), "Waste Not, Want Not? Water Tariffs for Sustainability", Report to WWF-UK, September 2007, available at *www.wwf.org.uk/filelibrary/pdf/water_tariffs_report01.pdf.*

Herrington, P. (2008), "Portuguese Proposals for Future Household Water Tariffs", report prepared for the 2007-08 OECD Horizontal Water Programme.

Hutton, G. and J. Bartram (2008), "Global Costs of Attaining the Millennium Development Goal for Water Supply and Sanitation", *Bulletin of the World Health Organization* 86(1), January.

INEGI (Instituto Nacional de Estadística, Geografía e Informática) (2003), "Censo de Captación, Tratamiento y Suministro de Agua", INEGI, Mexico.

IWA (International Water Association) (2008), "International Statistics for Water Services", IWA Statistics and Economics Specialist Group, presented at the IWA Water Congress, Vienna, available at *www.iwahq.org*.

Komives, K., *et al.* (2005), *Water, Electricity, and the Poor — Who Benefits from Utility Subsidies?*, The World Bank, Washington, DC.

Massarutto, A., *et al.* (2010), Financial Sustainability of Water and Sanitation Undertakings - An Application to Italy, Germany and the United States", OECD.

OECD (2003a), *Social Issues in the Provision and Pricing of Water Services*, OECD Publishing, DOI: *http://dx.doi.org/10.1787/9789264099890-en.*

OECD (2006a). "Water: The Experience in OECD Countries", *Environmental Performance Reviews, www.oecd.org/dataoecd/18/47/36225960.pdf.*

OECD (2006b), *Infrastructure to 2030: Telecom, Land Transport, Water and Electricity*, OECD Publishing, DOI: *http://dx.doi.org/10.1787/9789264023994-en.*

OECD (2007a), "The Health Costs of Inaction with Respect to Water Pollution", OECD unclassified document, ENV/EPOC/WPNEP(2006)17/FINAL, *www.olis.oecd.org/olis/2006doc.nsf/linkto/env-epoc-wpnep(2006)17-final.*

OECD (2007b), *OECD Environmental Performance Reviews: Switzerland 2007*, OECD Publishing, DOI: *http://dx.doi.org/10.1787/9789264030541-en.*

OECD (2007c), *OECD Environmental Performance Reviews: Belgium 2007*, OECD Publishing, DOI: *http://dx.doi.org/10.1787/9789264031128-en.*

OECD (2008a), "OECD Environmental Data Compendium", *www.oecd.org/document/49/0,3343,en_2649_34283_39011377_1_1_1_1,00.html.*

OECD (2008b), *OECD Environmental Outlook to 2030*, OECD Publishing, DOI: *http://dx.doi.org/10.1787/9789264040519-en.*

OECD (2008c), "Key Environmental Indicators 2008", *www.oecd.org/dataoecd/20/40/37551205.pdf.*

OECD (2008d), *OECD Environmental Performance Reviews: Denmark 2007*, OECD Publishing, DOI: 10.1787/9789264039582-en, *http://dx.doi.org/10.1787/9789264039582-en.*

OECD (2008e), *Growing Unequal?: Income Distribution and Poverty in OECD Countries*, OECD Publishing, DOI: *http://dx.doi.org/10.1787/9789264044197-en.*

Ofwat (Office of Water Services) (2009), "Future Water and Sewerage Charges 2010-15: Final Determinations", Ofwat, Birmingham, *www.ofwat.gov.uk/pricereview/pr09phase3/det_pr09_finalfull.pdf.*

Pires, J.S. (2007), "Consumer Tariffs in Practice: The Portuguese Experience", presentation made at the OECD Expert Meeting on Water Pricing and Financing, 14-15 November, available at *www.oecd.org/water*

Prüss-Üstün, A., *et al.* (2008), "Safer Water, Better Health: Costs, Benefits and Sustainability of Interventions to Protect and Promote Health", WHO, Geneva.

Smets, H. (2008), *De l'Eau Potable à un Prix Abordable: La Pratique des États*, V. Johanet, Paris, available at *www.academie-eau.org.*

UNESCO (United Nations Educational, Scientific and Cultural Organization) (2006), "United Nations World Water Development Report: Water, A Shared Responsibility", *WWDR No. 2,* World Water Assessment Programme, *www.unesco.org/water/wwap/wwdr/wwdr2.*

WHO/UNICEF (World Health Organization / United Nations Children's Fund) (2006), *Meeting the MDG Drinking Water and Sanitation Target: The Urban and Rural Challenge of the Decade*, UNICEF, New York and WHO, Geneva.

Annex A. Taxes and levies in household water bills

Table A.1. **Taxes and levies in household water and wastewater services**

	Water			Wastewater		
	VAT (%)	Other taxes	Abstraction charge	VAT (%)	Other taxes	Pollution charge
Belgium:						
Flanders	6.0	-	Y	-	No	No
Brussels	6.0	-	Y	6.0	No	Y
Wallonia	6.0	Y	Y	6.0	No	Y
Canada	-	-	Y	-	-	Y
Czech Republic	9.0	-	Y	9.0	No	n.a.
Denmark	25.0	Y	Y	25.0	Y	Y
Finland	22.0	-	No	22.0	-	No
France	5.5	Y	Y	5.5	Y	Y
Germany[1]	7.0	Y	Y	-	..	Y
Greece	9.0			19.0		
Hungary[2]	20.0		Y	20.0	-	Y
Italy	10.0	-	n.a.	10.0	No	n.a.
Japan	5.0			5.0		
Korea	-	Y	-	-	-	Y (partially)
Luxembourg	3.0			3.0		
Mexico	-	-	y	15.0	-	-
Netherlands	6.0			0.0		
New Zealand	-			-		
Norway	25.0			25.0		
Poland	7.0			7.0		
Portugal	5.0	-	Y	-	-	Y
Spain	7.0	-	Y	7.0	Y	Y
Sweden	25.0	-	-	25.0	..	No
Switzerland	2.4	-	Y	7.6	No	..
Turkey	8.0			8.0		
United Kingdom						
N. Ireland	-	-	-	-	-	-

Notes:

1. BGW, 2008.

2. OECD (2008), *OECD Environmental Performance Reviews: Hungary 2008*, OECD Publishing, DOI: *http://dx.doi.org/10.1787/9789264049284-en*.

Sources: OECD 2007-08 Survey and GWI survey, unless otherwise specified.

Compared with the situation presented in OECD's 1999 *The Price of Water: Trends in OECD Countries*, four more countries reported the application of value-added tax (VAT) on domestic water bills: Greece (9%), Luxembourg (3%,) Poland (7%) and Switzerland (2.4%). VAT rates only declined in Turkey (from 15% to 8%), but increased substantially in the Czech Republic (from 5% to 9%) and Hungary (from 12% to 20%), which thus joins the minority of OECD countries[1] charging 20-25% rates. At the other end of the spectrum, domestic bills in Canada, Korea, New Zealand and the United Kingdom are VAT-exempt, although in Northern Ireland VAT is charged for industrial customers. This is also the case in Mexico where industrial and commercial customer bills include a 15% VAT rate. Most countries[2] charge 5-10% rates, and in this group Italy and Spain posted small increases compared to the data published in OECD's 1999 *The Price of Water: Trends in OECD Countries.*

Similar country clusters characterise VAT on wastewater services, where there was an even more marked increase than for water services. Compared to OECD (1999), 17 countries reported VAT charges on wastewater bills as opposed to 10, the new ones being Belgium (6%, with the exception of the Flanders region), Greece (19%), Italy (10%), Luxembourg (3%), Mexico (15%), Poland and Spain (both at 7%) and Switzerland (7.6%). Hungary and Norway applied the same increase in VAT rates as for water bills, and the same decline took place in Turkey. In Portugal, VAT may be charged on wastewater services or not depending on the management model in the area, and currently it is paid by 70% of the population.

Table A.2. **Tariff structures for water supply and sanitation and policy objectives: a synthesis**

Tariff structure	Examples	Ecological sustainability - Encourage water savings - Improve water quality	Economic efficiency - Allocate to most beneficial uses - Avoid over-investment - Efficient use of existing facilities - Operational efficiency	Financial sustainability - Ensure viability of systems - Guarantee remuneration of inputs - Maintain asset values over time - Sustainable investment	Equity/affordability - Share costs in a fair way - Ensure affordability of merit users
Uniform flat fee	Some customers of some providers in the United Kingdom. Still used by many non-OECD utilities.	Very poor. No incentives for water saving nor for other aspects of sustainable water use.	Poor for drinking water (no linkage between fee structure and behaviour that may help minimise investment). OK for water services whose costs do not depend on water consumption.	Potentially OK, but commitment to cost recovery is what really matters. Need to avoid political determination of fees.	Very regressive (unless properly integrated with other elements of a social security system).
Non-uniform flat rate linked with households characteristics, *e.g.* property value or other income proxy, dwelling characteristics linked with water use	70% of UK households, common in the Former Soviet Union.	Poor if linked with income-related variable. Good if based on dwelling characteristics linked with: water use (*e.g.* use of water recycling devices, water-saving sprinklers, efficient appliances) or behaviour to be encouraged (*e.g.* rainwater harvesting, use of less pollutant detergents).	As above.	As above, provided that total revenues are guaranteed.	Potentially good criteria if linked with personal wealth. Regressive otherwise (unless properly integrated with other elements of a social security system).

Table A.2. **Tariff structures for water supply and sanitation and policy objectives: a synthesis** *(continued)*

Tariff structure	**Examples**	**Ecological sustainability** - Encourage water savings - Improve water quality	**Economic efficiency** - Allocate to most beneficial uses - Avoid over-investment - Efficient use of existing facilities - Operational efficiency	**Financial sustainability** - Ensure viability of systems - Guarantee remuneration of inputs - Maintain asset values over time - Sustainable investment	**Equity/affordability** - Share costs in a fair way - Ensure affordability of merit users
Uniform volumetric rate + 0 fixed charge	Numerous OECD countries. Most recurrent in sample of non-OECD utilities.	High, depending on level of volumetric rate. Higher than fixed + variable rate (all other things being equal, particularly revenue levels obtained) since 0 fixed charge means a larger marginal rate	Efficient if water is scarce or infrastructure nearing capacity or if variable costs high *vs.* fixed costs. Not very efficient otherwise: it would reduce societal benefits due to underutilisation of existing infrastructure.	Good potential for full cost recovery (FCR). Can have (temporary) negative impact on revenue in case of a sudden move from flat charges due to impact on demand (*e.g.* Berlin experience).	Depends on income elasticity. If this is low, it can hit large poor households hard.
Uniform volumetric rate + fixed charge > 0	Classic, *e.g.* Germany	High, depending on level of volumetric rate (impact on demand only if it is high enough) penetration of individual metering.	Optimal provided the following applies: volumetric rate = SRMC (short-run marginal cost) and fixed charge = recurrent fixed costs. Particularly suited if SRMC is constant (*e.g.* electricity, reagents).	As above.	Depends on size of fixed charge, but tends to be regressive (not so only if marginal cost is high and income elasticity is high – which is rare). Improved design: size of fixed charge can be differentiated based on income or proxies.

Table A.2. **Tariff structures for water supply and sanitation and policy objectives: a synthesis** *(continued)*

Tariff structure	Examples	**Ecological sustainability** - Encourage water savings - Improve water quality	**Economic efficiency** - Allocate to most beneficial uses - Avoid over-investment - Efficient use of existing facilities - Operational efficiency	**Financial sustainability** - Ensure viability of systems - Guarantee remuneration of inputs - Maintain asset values over time - Sustainable investment	**Equity/affordability** - Share costs in a fair way - Ensure affordability of merit users
Uniform volumetric rate + rebate (fixed charge < 0)	No known application. May have been applied in municipalities in the United States.	As above. Highest if rebates take into account specific circumstances (*e.g.* use of water recycling devices, water-saving sprinklers in gardens) or specific behaviour that wants to be encouraged (*e.g.* rainwater harvesting, use of less polluting detergents).	As above. In turn, could be efficient in combination with a positive fixed fee (idea: r = SRMC; fixed cost redistributed including a rebate for the poor).	As above.	Progressive and useful for reducing impact on poor. But only if rebate is targeted; otherwise, distributive effect depending on income elasticity, just like with IBTs.
Traditional IBT (both block widths and prices fixed) + fixed charge	Italy Increasing number of developing countries.	Highest, provided that metering is individual and marginal rates in the upper blocks are high.	Potentially the best solution provided r = SRMC and fixed charge = lump sum. Particularly suited in case SRMC is increasing (*e.g.* costly extra supply to be purchased).	Good potential for FCR if properly designed: lowest block not so large as to encompass most customers blocks structured so as to achieve target average tariff highest block not so high as to push largest users/sources of revenue out of the system.	Can be very regressive if: (i) larger poor households and low demand elasticity to income; (ii) average tariff is below cost recovery levels, discouraging network extensions; (iii) many households share the same tap.

Table A.2. **Tariff structures for water supply and sanitation and policy objectives: a synthesis** *(continued)*

Tariff structure	**Examples**	**Ecological sustainability** - Encourage water savings - Improve water quality	**Economic efficiency** - Allocate to most beneficial uses - Avoid over-investment - Efficient use of existing facilities - Operational efficiency	**Financial sustainability** - Ensure viability of systems - Guarantee remuneration of inputs - Maintain asset values over time - Sustainable investment	**Equity/affordability** - Share costs in a fair way - Ensure affordability of merit users
IBT + fixed charge + exact occupancy amendment	Flanders, Brussels Malta, some communes in Luxembourg	As above, but reduced incentives for large families.	Depends on how closely the resulting average volumetric charge reflects SRMC. Rest as above.	As above.	Reduces impact on large families (best if accompanied by reduction of leaks and improved efficiency of appliances). Depends on correlation of size and income of households. Problem (ii) above remains.
IBT + fixed charge + low-income households may apply for extension	Proposed Social Tariff Plan in Portugal	As above, but reduced incentives for low-income households that apply for extension of blocks.	Good for reducing demand in peak periods and optimising capacity use.	Uncertainty about number of households applying (may be reduced over time).	Very successful, if all eligible claim and block width reflect consumption patterns of the poor. Problem (ii) above remains.

Table A.2. **Tariff structures for water supply and sanitation and policy objectives: a synthesis** *(continued)*

Tariff structure	**Examples**	**Ecological sustainability** - Encourage water savings - Improve water quality	**Economic efficiency** - Allocate to most beneficial uses - Avoid over-investment - Efficient use of existing facilities - Operational efficiency	**Financial sustainability** - Ensure viability of systems - Guarantee remuneration of inputs - Maintain asset values over time - Sustainable investment	**Equity/affordability** - Share costs in a fair way - Ensure affordability of merit users
IBT + fixed charge + larger households (*e.g.* N>4) may apply for extension	Many Spanish cities. Greek DEYA, cities. Proposed option in Portugal.	As above, but reduced incentives for large families that apply for extension of blocks.	Depends on whether there is a fixed charge or not.	As above.	Depends on correlation of size and income of households. Problem (ii) above remains.
IBT + fixed charge + targeted subsidies to low income	Chile	Highest, provided that metering is individual and marginal rates in the upper blocks are high.	As above.	As above.	Depends on the capacity to target the poor. Problem (ii) above remains.

Table A.3. Comprehensive data on tariffs charged to households for water supply and sanitation in OECD countries[1]

Tariffs charged to households for water supply services in OECD countries

	Previous OECD surveys				New OECD estimates				Average annual rate of change %			GWI (2008)
	OECD (1999)		OECD (2003)									
	Year	USD	Year	USD	Year	Measure	USD	PPPs	Period	Nominal	Real	USD[2]
Australia	1996-97	0.74	2000	0.67	2007	WAV (250)	1.15	0.94	2003-07	5.49	2.69	1.59
Austria	1997	1.06	1999	1.12	2000	(a)	0.99	1.17	..	..	..	2.19
Belgium	..	..	1999	1.68					..	..	..	4.90
Flanders	1997	1.68	1997	1.68	2004	Median	1.66	1.45	..	..	..	..
Brussels	1997	1.67	1997	1.67	2007	AV(180)	2.48[5]	1.98	..	..	..	..
Wallonia	1997	1.69	1997	1.69	2006	AVE (100)	2.42[4,6]	2.09	2005-06[4]	3.80	1.98	..
Canada	..	..	1999	0.46	2004	WAV (300): 70%	0.52[4]	0.53	1999-04[4]	1.57	-0.32	1.39
Czech Republic	1997	0.38	2000	0.54	2007	AV	1.31[4]	1.75	2000-07[4]	5.79	3.31	1.61
Denmark	1995	1.34	2001	1.90	..	..	..	..	..	..	..	8.61
Finland	1998	1.29	2002	1.16	2008	(a)	1.72[4]	1.09	2000-08[3]	2.83	1.26	1.60
France	1995	1.50	2000	1.16	2005	WAV (120)	1.86	1.60	2000-05	1.98	0.07	3.96
Germany	1997	1.69	2001	1.52	2007	AV	2.53	2.06	2000-07	1.30	-0.63	3.07
Greece	1995	0.81	2002	0.56	2006	AV	0.95	0.98	2000-06	2.38	-0.96	0.94
Hungary	1997	0.39	2002	0.61	2005	AV	1.05	1.52	2000-05	8.66	2.65	1.25
Iceland	..	..	n.a.	n.a.	..		..	..	..	..	..	1.43
Ireland	..	..	..	..	..		0.00	..	..	..	..	0.00
Italy	1996	0.51	2001	0.39	2007	(a)	0.82	0.66	2005-07	4.45	2.44	0.94
Japan	1995	1.50	2001	1.19	2003	(a)	..	..	1999-03	n.a	n.a	1.05
Korea	1996	0.25	n.a.	n.a.	2006	AV	0.61[4]	0.66	2000-06[3]	4.42	1.23	0.55
Luxembourg	1994	1.08	1994	1.08	1999	(a)	1.03	1.06	1994-99	1.62	0.34	3.55
Mexico	..	..	2001	0.25	..		..	..	..	..	..	0.33
Netherlands	1998	1.46	2000	1.45	2007	WAV (55%)	2.29	1.85	2000-07	0.89	-1.33	2.24
New Zealand	..	..	n.a.	n.a.	2007	AB (250)	0.74	0.63	2003-07	-3.52	-6.11	1.28

Tariffs charged to households for water supply services in OECD countries *(continued)*

	Previous OECD surveys				New OECD estimates				Average annual rate of change			GWI
	OECD (1999)		OECD (2003)							%		(2008)
	Year	USD	Year	USD	Year	Measure	USD	PPPs	Period	Nominal	Real	USD[2]
Norway	..	..	..	..	..		..	..	..	..	..	..
Poland	..	..	2002	0.52	2007	(a)	0.93	1.22	..	..	..	1.28
Portugal	..	..	2002	0.50	2007	WAVE(200):1 00%	1.23[4]	1.20	2004-07[3]	2.74	0.00	1.31
Slovak Republic	..	..	n.a.	n.a.	..		..	..	..	..	..	1.49
Spain	1994	0.72	n.a.	n.a.	2006	(a)	0.89[4]	0.87	2000-06[3]	4.03	0.74	1.17
Sweden	1997	1.09	1999	1.15	..	(a)	..	..	..	..	..	2.15
Switzerland	1996	1.29	1996	1.29	2007	AB (180)	1.20[4]	0.78	..	..	..	2.72
Turkey	1998	1.01	1998	1.01	..		..	..	..	..	..	1.77
United Kingdom:									..	..	..	2.21
England and Wales	1998-99	1.42	2001	1.12	2006	AB (137)	1.79	1.48	2001-06	4.45	2.73	2.57
N. Ireland	..	..	..	..	..	AV(180)	..	..	..	..	..	..
Scotland	1997-98	0.84	n.a	n.a	2007	AB (110)	2.60	2.00	2004-07	2.65	0.41	2.93
United States	1997	0.58	2001	0.63	..		..	..	..	..	..	0.93

1. Based on official OECD USD and PPP-adjusted exchange rates with reference to the year of the data point.
2. The USD exchange rates applied by GWI may not be the same used in the OECD surveys. This may affect comparisons.
3. Data provided by the country (OECD 2007-08 Survey). No additional computation by OECD.
4. OECD estimates based on data provided by the country (OECD 2007-08 Survey).
5. OECD estimates based on publicly available data. Validation by the country is pending.
6. Belgium Wall.: VAT and Fond Social excluded.

Measures used:

(a)	not specified.
AB(x)	Average bill covering flat-rate or metered customers, with a representative household consumption of x m^3/year.
WAVB(x)	Weighted average bill, with household consumption of x m^3/year and utilities in the sample covering P% of population.
AV	Average price of water across all utilities and households = Total revenue from households/Total m^3 sold.
AVE(x): P	Average volumetric rates across utilities plus a "volumetric equivalent" average fixed charge, with an assumed average household consumption of x m^3/year and utilities in the sample covering P% of the country's population.
WAV(x): P	Weighted averages of volumetric rates, with an assumed average household consumption of x m3/year and utilities in the sample covering P% of the population.
WAVE(x): P	Weighted average of volumetric rates and a "volumetric equivalent" average fixed charge, with an assumed average household consumption of x m^3/year and utilities in the sample covering P% of the population

Tariffs charged to households for sanitation services in OECD countries[1]

	Previous OECD surveys				New OECD estimates				Average annual rate of change %			GWI (2008)
	OECD (1999)		OECD (2003)									
	Year	USD	Year	USD	Year	Measure	USD	PPPs	Period	Nominal	Real	USD[2]
Australia	1996-97	0.87	2000	0.83	2007	WAV (250)	1.29	1.05	2003-07	4.6	1.9	1.81
Austria	1997	..	2003	1.53			..	..	..	..	..	1.63
Belgium												0.11
Flanders	1997	0.69	1997	0.69	2004	Median	0.82	0.71	..	..	..	..
Brussels	1997	0.39	1997	0.39	2007	AV(180)	1.10[5]	0.88	..	..	..	..
Wallonia	1997	0.45	1997	0.45	2006	AVE (100)	0.78[4,6]	0.68	2005-06[4]	19.5	17.4	..
Canada	1994	..	1999	0.32	2004	WAV (300): 70%	0.45[4]	0.46	1999-04[5]	7.6	5.6	0.62
Czech Republic	1997	0.30	2000	0.54	2007	AV	1.12[4]	1.50	2000-07[4]	6.0	1.3	1.47
Denmark	1995	1.84	2001	2.20			..	..	2000-06			0
Finland	1998	1.57	2002	1.46	2008	(a)	2.68[4]	1.69	2000-08[3]	3.9	2.3	2.03
France	1995	1.50	2000	1.27	2005	WAV(120)	1.99	1.71	2000-05	6.3	4.3	0.39
Germany	1997	..	2001	1.95			..	..	..	..	..	2.49
Greece	1995	0.32	2002	0.28	2006	AV	0.45	0.46	2000-06	2.8	-0.5	0.64
Hungary	1997	0.28	2002	0.49	2005	AV	0.97	1.41	2000-05	12.0	5.8	1.59
Iceland	..	..	..	..			..	..	..	..	..	0
Ireland	..	..	..	..	2007		0.00	..	..	..	..	0
Italy	1996	0.34	2001	0.21	2007	(a)	0.70	0.56	2005-07	6.5	4.4	0.33
Japan	1995	1.13	..	..	2003	(a)	..	..	..	..	..	0.81
Korea	1996	0.11	..	..	2006	AV	0.25[4]	0.28	2000-06[3]	10.8	7.4	0.18
Luxembourg	1994	..	..	..			..	..	..	..	..	2.11
Mexico	..	..	..	..			..	..	..	..	..	0.04
Netherlands	1998	1.81	2000	1.68	2006	WAV (55%)	..	..	..	..	..	0.32
New Zealand	..	..	..	..	2007	AB (250)	1.24	1.06	2003-07	-3.9	-6.5	2.39
Norway	..	..	..	..			..	..	..	..	..	..

Tariffs charged to households for sanitation services in OECD countries[1] *(continued)*

	Previous OECD surveys				New OECD estimates				Average annual rate of change			GWI
	OECD (1999)		OECD (2003)						%			(2008)
	Year	USD	Year	USD	Year	Measure	USD	PPPs	Period	Nominal	Real	USD[2]
Poland	..	..	2002	0.58	2007	(a)	1.19	1.56	..	..	..	1.44
Portugal	..	..	..	..	2007	WAVE(200):100%	0.47[4]	0.46	2004-07[3]	2.2	0.0	0.58
Slovak Republic	..	..	..	..					..	..	..	1.23
Spain	1994	0.37	..	..	2006	(a)	0.46[4]	0.46	2000-06[5]	13.8	10.2	0.84
Sweden	1997	1.62	1999	1.71			..	..	..	..	..	0.22
Switzerland	1996	..	..	..	2007	AB (180)	1.93[4]	1.26	..	..	..	2.78
Turkey	1998	0.51	1998	0.51			..	..	..	..	..	0.32
United Kingdom	..	..	..	..			..	..	..	..	..	1.71
England and Wales	1998-99	1.67	2001	1.26	2006	AB(137)	2.03	1.68	2001-06	4.7	3.0	3.27
N. Ireland	..	..	..	..			..	..	..	..	..	..
Scotland	1997-98	0.59	..	..	2007	AB (110)	3.11	2.39	2004-07	2.6	0.4	3.78
United States	1997	0.67	2001	0.77	..		..	..	..	..	..	1.17

1. Based on official OECD USD and PPP-adjusted exchange rates with reference to the year of the data point.
2. The USD exchange rates applied by GWI may not be the same used in the OECD surveys. This may affect comparisons.
3. Data provided by the country (OECD 2007-08 Survey). No additional computation by OECD.
4. OECD estimates based on data provided by the country (OECD 2007-08 Survey).
5. OECD estimates based on publicly available data. Validation by the country is pending.
6. Belgium Wall.: VAT and Fond Social excluded.

Measures used:

(a)	not specified.
AB(x)	Average bill covering flat-rate or metered customers, with a representative household consumption of x m^3/year.
AV	Average price of water across all utilities and households = Total revenue from households/Total m^3 sold.
WAVB(x)	Weighted Average bill of utilities, with household consumption of x m^3/year.
AVE(x): P	Average of volumetric rates across utilities plus a "volumetric equivalent" average fixed charge, with an assumed average household consumption of x m^3/year and (when available) utilities in the sample covering P% of the country's population.
WAV(x): P	Weighted averages of volumetric rates, with an assumed average household consumption of x m3/year and utilities in the sample covering P% of the population.
WAVE(x): P	Weighted averages of volumetric rates and a "volumetric equivalent" average fixed charge, with an assumed average household consumption of x m^3/year and utilities in the sample covering P% of the population

Source: OECD, GWI, IWA.

Tariffs charged to households for water supply and sanitation services (total) in OECD countries[1]

	Previous OECD surveys				New OECD estimates based on						
	OECD (1999)		OECD (2003)		W and WW tariffs (see notes)				OECD 2008 Survey AB (180)		GWI (2008) AB (180)
	Year	USD	Year	USD	Year	Measure	USD	PPPs	USD	PPPs	USD[2]
Australia	1996-97	1.61	2000	1.5	2007	WAV (250)	2.44	1.99	..	..	3.40
Austria	1997	..	1999	2.6			..	..	..	..	3.81
Belgium											5.01
Flanders	1997	2.36	1997	2.3	2004	Median	2.48	2.16	4.14	3.27	..
Brussels	1997	2.06	1997	2.0	2007	AVE (180)	3.58[5]	2.85	..	..	..
Wallonia	1997	2.14	1997	2.1	2006	AVE (100)	3.20[3,6]	2.76	3.92	3.10	..
Canada	1994	0.70	1999	0.7	2004	WAV (300): 70%	0.97[4]	0.99	1.58	1.66	2.01
Czech Republic	1997	0.68	2000	0.9	2007	AV	2.43[3]	3.25	..	..	3.08
Denmark	1995	3.18	2001	4.1	2006	(a)	6.7	4.41	..	..	8.61
Finland	1998	2.86	2002	2.6	2008	(a)	4.41[3]	2.78	..	..	3.63
France	1995	3.00	2000	2.4	2005	WAV (120)	3.85	3.30	3.74	3.16	4.34
Germany	1997	..	2001	3.4			..	..	..	..	5.56
Greece	1995	1.14	2002	0.8	2006	AV	1.4[5]	1.45	..	..	1.58
Hungary	1997	0.67	2002	1.1	2005	AV	2.02[5]	2.93	..	..	2.84
Iceland		..	..	..			..	..	..	..	1.43
Ireland		..	..	..			..	..	..	..	0.00
Italy	1996	0.84	2001	0.6	2007	(a)	1.52	1.23	1.45	1.20	1.27
Japan	1995	2.63	..	..	2003	(a)	1.85[5]	1.38	..	..	1.86
Korea	1996	0.36	..	..	2006	AV	0.86[3]	0.94	0.77	0.82	0.73
Luxembourg	1994	..	..	..			..	..	..	..	5.67
Mexico		..	..	..	2007	AVE (180): 18%	0.43	0.62	0.49	0.69	0.37
Netherlands	1998	3.27	2000	3.1			..	..	..	..	2.56
New Zealand		..	..	..	2007	AB (250)	1.98[5]	1.69	..	..	3.67
Norway		..	..	..			..	..	..	..	3.76

Tariffs charged to households for water supply and sanitation services (total) in OECD countries[1] *(continued)*

	Previous OECD surveys				New OECD estimates based on						
	OECD (1999)		OECD (2003)		W and WW tariffs (see notes)				OECD 2008 Survey AB (180)		GWI (2008) AB (180)
	Year	USD	Year	USD	Year	Measure	USD	PPPs	USD	PPPs	USD[2]
Poland		..	2002	1.1	2007	(a)	2.12[5]	2.79	..	..	2.72
Portugal		..	..	..	2007	WAVE(200):100%	1.70[3]	1.67	1.23	1.20	1.89
Slovak Republic		..	..	..			..	..	..	..	2.72
Spain	1994	1.09	..	..	2006	(a)	1.36[3]	1.33	1.92	1.89	2.02
Sweden	1997	2.71	1999	2.8	2008	(a)	3.59[3]	2.82	3.59	2.82	2.37
Switzerland	..	..	..	..	2007	AB (180)	3.13[4]	2.05	3.13	2.05	5.50
Turkey	1998	..	..	..			..	..	..	..	2.09
United Kingdom									..	..	3.92
England and Wales	1998-89	3.10	2001	2.3	2006	AB (137)	3.82	3.15	..	..	5.84
N. Ireland		..	..	..			..	..	..	..	..
Scotland	1997-98	1.43	2000	2.3	2007	AB (110)	5.72	4.39	..	..	6.71
United States	1997	1.25	2001	1.4			..	..	..	..	2.10

1. Based on official OECD USD and PPP-adjusted exchange rates with reference to the year of the data point.
2. The USD exchange rates applied by GWI may not be the same used in the OECD surveys. This may affect comparisons.
3. Data provided by the country (OECD 2007-08 Survey). No additional computation by OECD.
4. OECD estimates based on data provided by the country (OECD 2007-08 Survey).
5. OECD estimates based on publicly available data. Validation by the country is pending.
6. Belgium Wall.: VAT and Fond Social excluded.

Measures used:

(a) not specified.
AB(x) Average bill covering flat-rate or metered customers, with a representative household consumption of x m^3/year.
AV Average price of water across all utilities and households = Total revenue from households/Total m^3 sold.
AVB(x) Average bill of utilities, with household consumption of x m^3/year.
AVE(x): P Average of volumetric rates plus a "volumetric equivalent" average fixed charge, with an assumed average household consumption of x m^3/year and utilities in the sample covering P% of the country's population.
WAV(x): P Weighted averages of volumetric rates, with an assumed average household consumption of x m3/year and utilities in the sample covering P% of the population.
WAVE(x): P Weighted averages of volumetric rates and a "volumetric equivalent" average fixed charge, with an assumed average household consumption of x m^3/year and utilities in the sample covering P% of the population.

Notes

1. The others are Denmark, Finland, Sweden and Norway, which increased its rate from 22% to 25%.
2. Belgium, the Czech Republic, France, Germany, Greece, Japan, the Netherlands, Poland, Portugal, Spain, Turkey.

Annex B. Additional sources of information per country

Australia

WSAA (Water Services Association of Australia, Inc.) (2003), "WSAAfacts 2003", Melbourne.

WSAA (2004), "WSAAfacts 2004", Melbourne.

WSAA (2005), "WSAAfacts 2005", Melbourne.

WSAA (2007), "National Performance Report 2005-06 Major Urban Water Utilities", Melbourne.

WSAA (2008), "National Performance Report 2005-06 Major Urban Water Utilities", Melbourne.

Austria

OVGW (2002), "Betriebsergebnisse der Wasserwerke Österreichs" , available at *www.ovgw.at/en/themen/index_html?uid:int=176.*

Belgium

Brussels: IBDE (2007), *www.ibde.be/index.cfm?Content_ID=892891470.*

Wallonia: AQUAVAL (2006), "Chiffres-clés du secteur de l'eau en Région Wallonne -Année 2006", *www.aquawal.be/xml/liste-IDC-249-.html.*

Flanders: OECD (2007), *OECD Environmental Performance Reviews: Belgium 2007*, OECD Publishing, DOI: *http://dx.doi.org/10.1787/9789264031128-en.*

Canada

Environment Canada (2001), "2001 Municipal Water Pricing (MUP) Summary Tables", available at *www.ec.gc.ca/water/en/info/pubs/sss/PricingTables2001.pdf.*

Environment Canada (2004), "2004 Municipal Water Pricing (MUP) Summary Tables", available at *www.ec.gc.ca/water/fr/info/pubs/sss/TableauPrix2004.pdf.*

Czech Republic

Czso (2007), "Water Supply, Sewerage and Watercourses in 2007", CZSO database, available at *www.czso.cz/csu/2008edicniplan.nsf/engpubl/2003-08-in_2007.*

Denmark

DANVA (2007), "Statistik udviklingstabeller og figurer med data for 2006, Figurer fra Vandstatistikken", *www.bessy.dk/java/viewReports.do?year=2007.*

Finland

Finnish Water and Wastewater Works Association VVY (2007), "VESIHUOLTOMAKSUT1.2.2007 - VVY: N JÄSENLAITOKSET", available at *www.vvy.fi.*

Ministry of Environment (2008), "Detailed Information on Environment-related Taxes and Charges in Finland", *www.ymparisto.fi/download.asp?contentid=85625&lan=en.*

France

IFEN (2005), "L'économie de l'environnement en 2003", available at *www.ifen.fr/publications/DonEco/doneco2005/comptes2003.pdf.*

IFEN (2007), "Rapport de la Commission des Comptes et de l'Économie de l'Environnement, 2005", available at *www.ifen.fr/publications/syntheses/PDF/comptes2007.pdf.*

Germany

BGW (2005), "Profile of the German Water Industry, available at *www.dvgw.org/pdf/branchenbild_en.pdf.*

BGW (2008), "Profile of the German Water Industry", available at *www.dvgw.de/fileadmin/dvgw/wasser/organisation/branchenbild2008_en.pdf.*

Greece

EYDAP (2006), "Annual Bulletin, 2006" available at *www.eydap.gr/media/financialdata/annual_reports_en/annual_bulletin_2006.pdf.*

Italy

Osservatorio Prezzi-Cittadinanza Attiva (2006), "Indagine svolta dall'Osservatorio Prezzi & Tariffe sulle tariffe idriche in Italia", *www.tonix.it/Cittadinanzattiva_servizio_idrico_2007.pdf.*

Osservatorio Prezzi-Cittadinanza Attiva (2008), "Indagine svolta dall'Osservatorio Prezzi & Tariffe sulle tariffe idriche in Italia", *www.cittadinanzattiva.it/files/consumatori/DOSSIER_ACQUA_2008.pdf.*

Korea

Ministry of Environment (2006), "Water Quality Management 2006", available at *http://eng.me.go.kr/docs/common/common_view.html?idx=15&av_pg=1&mcode=20&classno=13.*

Luxembourg

ALUSEAU (2003), "Année Internationale de l'eau douce (2003) ", available at *www.aluseau.lu/Prix%20de%20l%27eau%20pratiqu%E9s%20au%20Luxbg.pdf.*

Ministère de l'Intérieur et de l'Aménagement du Territoire (2006), " Rapport d'activité 2006", available at *www.gouvernement.lu/publications/informations_gouvernementales/rapports_activite/rapport_activite2006/13int/miat.pdf.*

Mexico

CNA (Comisión Nacional del Agua) (2002), "Situación del Subsector Agua Potable, Alcantarillado y Saneamiento a Diciembre de 2001", Mexico.

CNA (2003), "Situación del Subsector Agua Potable, Alcantarillado y Saneamiento a Diciembre de 2002", Mexico.

CNA (2004), "Situación del Subsector Agua Potable, Alcantarillado y Saneamiento a Diciembre de 2003", Mexico.

CNA (2005), "Situación del Subsector Agua Potable, Alcantarillado y Saneamiento a Diciembre de 2004", Mexico.

CNA (2006), "Situación del Subsector Agua Potable, Alcantarillado y Saneamiento a Diciembre de 2005", Mexico.

CNA (2007), "Estadísticas del Agua en México", Mexico.

New Zealand

WSAA (Water Services Association of Australia, Inc.) (2003), "WSAAfacts 2003", Melbourne.

WSAA (2004), "WSAAfacts 2004", Melbourne.

WSAA (2005), "WSAAfacts 2005", Melbourne.

WSAA (2007), "National Performance Report 2005-06 Major Urban Water Utilities", Melbourne.

WSAA (2008), "National Performance Report 2005-06 Major Urban Water Utilities", Melbourne.

Netherlands

VEWIN (2001), "Water Supply Statistics 2001", Vewin: Rijswijk, available at *www.vewin.nl/bestanden/internet/Publicaties/RappDocs/2002/Water_Supply_Statistics_2001.pdf.*

VEWIN (2002), "Water Supply Statistics 2002", Vewin: Rijswijk, available at *www.vewin.nl/bestanden/internet/Publicaties/RappDocs/2003/Water_Supply_Statistics_2002.pdf.*

VEWIN (2003), "Water Supply Statistics 2003, Vewin: Rijswijk, available at *www.vewin.nl/bestanden/internet/Publicaties/RappDocs/2004/Water_Supply_Statistics_2003.pdf.*

VEWIN (2004), "Water Supply Statistics 2004, Vewin: Rijswijk, available at *www.vewin.nl//Water%20supply%20Statistics%202004.pdf.*

VEWIN (2005), "Water Supply Statistics 2005, Vewin: Rijswijk, available at *www.vewin.nl//bestanden/internet/Publicaties/RappDocs/2006/WSS%20eng%202005%20web.pdf.*

VEWIN (2006), "Water Supply Statistics 2006, Vewin: Rijswijk, available at *www.vewin.nl/bestanden/internet/Publicaties/VEWINuitgaven/Water%20Supply%20Statistics%202006%20webversie.pdf.*

VEWIN (2007), "Water Supply Statistics 2007, Vewin: Rijswijk, available at *www.vewin.nl/SiteCollectionDocuments/Publicaties/Overige%20Vewin-uitgaven/2008/Water%20Supply%20Statistics%202007.pdf.*

Norway

Statistics Norway (2007), KOSTRA database, available at *http://statbank.ssb.no/statistikkbanken/Default_FR.asp?PXSid=0&nvl=true&PLanguage=1&tilside=selecttable/hovedtabellHjem.asp&KortnavnWeb=vann_kostra.*

Poland

Central Statistical Office (2006), "Housing Economy in 2006", (data), available at *www.stat.gov.pl/gus/45_1956_ENG_HTML.htm.*

IGWP (2007), Polish Chamber for Water Economy, *www.igwp.info.pl.*

Sweden

Svenskt Vatten (2000), VA-taxor 2000. Sammanställning över kommunala vatten- och avloppstaxor gällande den 1 januari 2000, available at *www.svensktvatten.se/Templates/FileArchive1.aspx?PageID=50fe9d4d-1a5d-4a12-ae22-510302ebbcdb.*

Svenskt Vatten (2001), VA-taxor 2001. Sammanställning över kommunala vatten- och avloppstaxor gällande den 1 januari 2000, available at *www.svensktvatten.se/Templates/FileArchive1.aspx?PageID=50fe9d4d-1a5d-4a12-ae22-510302ebbcdb.*

Svenskt Vatten (2002), VA-taxor 2002. Sammanställning över kommunala vatten- och avloppstaxor gällande den 1 januari 2000, available at *www.svensktvatten.se/Templates/FileArchive1.aspx?PageID=50fe9d4d-1a5d-4a12-ae22-510302ebbcdb.*

Svenskt Vatten (2003), Taxestatistik 2003 - Sammanställning över kommunala vatten- och avloppstaxor gällande den 1 januari 2003, Rapporten är under utveckling, Sweden. , available at *www.svensktvatten.se/Templates/FileArchive1.aspx?PageID=50fe9d4d-1a5d-4a12-ae22-510302ebbcdb.*

Svenskt Vatten (2004), Taxestatistik 2004 - Sammanställning över kommunala vatten- och avloppstaxor gällande den 1 januari 2004, Rapporten är under utveckling, Sweden. *www.svensktvatten.se/Templates/FileArchive1.aspx?PageID=03505973-5c9d-4a3c-a372-5a4eb65b9c87.*

Svenskt Vatten (2005), Taxestatistik 2005 - Sammanställning över kommunala vatten- och avloppstaxor gällande den 1 januari 2005, Rapporten är under utveckling, Sweden. *www.svensktvatten.se/Templates/FileArchive1.aspx?PageID=a49f4401-073b-402c-a3ea-21557bf8b214.*

Svenskt Vatten (2006), Taxestatistik 2006 - Sammanställning över kommunala vatten- och avloppstaxor gällande den 1 januari 2006, Rapporten är under utveckling, Sweden. *www.svensktvatten.se/Templates/FileArchive1.aspx?PageID=fc29a2a7-6f9b-40cb-a278-13635ec3ee0c.*

Svenskt Vatten (2007), Taxestatistik 2007 - Sammanställning över kommunala vatten- och avloppstaxor gällande den 1 januari 2007, Rapporten är under utveckling, Sweden. *www.svensktvatten.se/Templates/FileArchive1.aspx?PageID=339eba27-f64e-40bf-a2fe-48a5a2d4e628.*

Svenskt Vatten (2008), Taxestatistik 2007 - Sammanställning över kommunala vatten- och avloppstaxor gällande den 1 januari 2007, Rapporten är under utveckling.

United Kingdom

Ofwat (Office of Water Services) (2002), "Tariff Structure and Charges: 2002-2003 Report", Ofwat, Birmingham, *www.ofwat.gov.uk/aptrix/ofwat/publish.nsf/AttachmentsByTitle/tariff_report03.pdf/$FILE/tariff_report03.pdf.*

Ofwat (2003), "Tariff Structure and Charges: 2003-2004 Report", Ofwat, Birmingham, *www.ofwat.gov.uk/aptrix/ofwat/publish.nsf/AttachmentsByTitle/tariff_report04.pdf/$FILE/tariff_report04.pdf.*

Ofwat (2004), "Tariff Structure and Charges: 2004-2005 Report", Ofwat, Birmingham, *www.ofwat.gov.uk/aptrix/ofwat/publish.nsf/AttachmentsByTitle/tariff_report05.pdf/$FILE/tariff_report05.pdf.*

Ofwat (2005), "Tariff Structure and Charges: 2005-2006 Report", Ofwat, Birmingham, *www.ofwat.gov.uk/aptrix/ofwat/publish.nsf/AttachmentsByTitle/tariff_report06.pdf/$FILE/tariff_report06.pdf.*

Ofwat (2006), "Tariff Structure and Charges: 2007-2008 Report", Ofwat, Birmingham, *www.ofwat.gov.uk/aptrix/ofwat/publish.nsf/AttachmentsByTitle/tariff_report07.pdf/$FILE/tariff_report07.pdf.*

Ofwat (2009), "Future Water and Sewerage Charges 2010-15: Final Determinations", Ofwat, Birmingham, *www.ofwat.gov.uk/pricereview/pr09phase3/det_pr09_finalfull.pdf.*

Scottish Water (2004), "Scheme of Charges 2004/5", available at *www.scottishwater.co.uk/portal/page/portal/SWE_PGP_ABOUT_US/SWE_PGE_ABOUT_US/SWE_PAGE2_SITE_HELP/SWE_PAGE3_SITE_HELP_SEARCH.*

Scottish Water (2006), "Scheme of Charges 2006/7", available at *www.scottishwater.co.uk/portal/page/portal/SWE_PGP_ABOUT_US/SWE_PGE_ABOUT_US/SWE_PAGE2_SITE_HELP/SWE_PAGE3_SITE_HELP_SEARCH.*

Scottish Water (2008), "Scheme of Charges 2007/8", available at www.scottishwater.co.uk/portal/page/portal/SWE_PGP_HOUSEHOLD/SWE_PGE_HOUSEHOLD/SWE_HH_CHRG/Copy_of_SWE_HH_CHRG_1_CHRG0708/Copy_of_SW_HH_CHARGES0708_MET.

OECD PUBLISHING, 2, rue André-Pascal, 75775 PARIS CEDEX 16
PRINTED IN FRANCE
(97 2010 04 1 P) ISBN 978-92-64-08346-2– No. 57289 2010